RECEIVED
RESEARCH LIBRARY

JUL 2 5

SCHERING PLOUGH

Electrically Assisted Transdermal and Topical Drug Delivery

New and Forthcoming Titles in the Pharmaceutical Sciences

Physiological Pharmaceutics Barriers to Drug Absorption
2nd Edition *Washington & Wilson* 1998 0 7484 0562 3 Hbk 0 7484 0610 7 Pbk

Intelligent Software for Product Formulation
Rowe & Roberts 1998 0 7484 0732 4 Hbk

Flow Injection Analysis of Pharmaceuticals
Automation in the Laboratory *Martínez Calatayud* 1996 0 7484 0445 7 Hbk

Photostability of Drugs and Drug Formulations
Tønneson (Ed) 1996 0 7484 0449 X Hbk

Pharmaceutical Coating Technology *Cole/Hogan/Aulton* 1995 013 6628915 Hbk

Dielectric Analysis of Pharmaceutical Systems
Craig 1995 0 13 210279 X Hbk

Microbial Quality Assurance in Cosmetics, Toiletries and Non-Sterile Pharmaceuticals 2nd Edition *Baird with Bloomfield (Eds)* 1996 0 7484 0437 6 Hbk

Immunoassay A Practical Guide *Law (Ed)* 1996 0 7484 0560 7 Hbk

Cytochromes P450 Structure, Function and Mechanism *Lewis* 1996 0 7484 0443 0 Hbk

Autonomic Pharmacology *Broadley* 1996 0 7484 0556 9 Hbk

Pharmaceutical Experimental Design and Interpretation 2nd Edition
Armstrong & James 1996 0 7484 0436 8 Hbk

Pharmaceutical Production Facilities 2nd Edition
Cole 1998 0 7484 0438 4 Hbk

full pharmaceutical science catalogue available
or
visit our new website on: http://www.tandf.co.uk

1900 Frost Road
Suite 101, Bristol PA,
19007 - 1598, USA
Tel: 1-800-821-8312
Fax: 215-785-5515

Rankine Road
Basingstoke, Hants,
RG24 8PR, UK
Tel: +44 (0)1256 813000
Fax: +44 (0)1256 479438

Electrically Assisted Transdermal and Topical Drug Delivery

AJAY K. BANGA, PH.D.

UK   Taylor & Francis Ltd, 1 Gunpowder Square, London EC4A 3DE
USA   Taylor & Francis Inc., 1900 Frost Road, Suite 101, Bristol, PA 19007

Copyright © Ajay. K. Banga 1998
All rights reserved. No part of this publication may be reproduced, stored in a retrieval system, or transmitted, in any form or by any means, electronic, electrostatic, magnetic tape, mechanical, photocopying, recording or otherwise, without the prior permission of the copyright owner.

British Library Cataloguing in Publication Data
A catalogue record for this book is available from the British Library.

ISBN 0-7484-0687-5

Library of Congress Cataloguing in Publication data are available

Typeset in 10/12pt Times by Graphicraft Typesetters Ltd, Hong Kong

Printed by T.J. International Ltd, Padstow, UK

To Saveta, my wife, for her understanding and encouragement

To Manisha, my toddler daughter, for her unconditional love

To my parents, for their moral support and encouragement

Contents

Preface	*Page* ix
About the author	xi

1 Percutaneous absorption and its enhancement   1
    1.1   Introduction   1
    1.2   Structure, enzymatic activity and immunology of the skin   2
    1.3   Percutaneous absorption   5
    1.4   Iontophoretic enhancement of percutaneous absorption   7
    1.5   Other methods for enhancement of percutaneous absorption   10

2 Iontophoretic transdermal drug delivery   13
    2.1   Introduction   13
    2.2   Electrical properties of the skin   13
    2.3   Pathways of iontophoretic drug delivery   15
    2.4   Theoretical basis of iontophoresis   16
    2.5   Factors affecting iontophoretic delivery   18
    2.6   Electro-osmotic flow   24
    2.7   Reverse iontophoresis   26
    2.8   Case studies   28

3 *In vitro* experimental techniques for iontophoresis research in the laboratory   33
    3.1   Introduction   33
    3.2   Selection of skin or membrane   33
    3.3   *In vitro* transdermal iontophoresis studies   37
    3.4   pH control in iontophoresis research   43
    3.5   Novel formulations for iontophoretic delivery   48

4 Clinical applications of iontophoresis devices for topical dermatological delivery   57
    4.1   Introduction   57
    4.2   Clinical applications of iontophoresis   57

	4.3	Iontophoresis devices	68
	4.4	Electrode design	71
	4.5	Treatment protocols and formulations	73

5 Iontophoretic delivery of peptides 75
5.1 Introduction 75
5.2 Structural considerations of polypeptides relevant to iontophoretic delivery 77
5.3 Peptides as iontophoresis candidates 78
5.4 Protease inhibitors and permeation enhancers for iontophoretic delivery 79
5.5 Case studies 82
5.6 Iontophoretic delivery of insulin: fact or fiction? 90

6 Transdermal delivery by electroporation 95
6.1 Introduction 95
6.2 Electropermeabilization of the skin 97
6.3 Electroporation equipment 98
6.4 Factors affecting delivery by electroporation 100
6.5 Pulsatile delivery of drugs by electroporation 102
6.6 Skin toxicology of electroporation 103
6.7 Delivery of particulates into the skin 105
6.8 Electroporation for cancer chemotherapy 106
6.9 Commercial development of electroporation 107

7 Electrically assisted delivery of gene-based drugs to skin 109
7.1 Introduction 109
7.2 Delivery of oligonucleotides 110
7.3 Delivery of genes 112
7.4 DNA vaccines 115

8 Developmental issues and commercialization of wearable iontophoretic patches 119
8.1 Introduction 119
8.2 Human studies 123
8.3 Dose and bioavailability 125
8.4 Biological and safety issues 126
8.5 Skin damage and its measurement 128
8.6 Patch design and manufacture 130
8.7 Developmental and regulatory issues 132

Bibliography 135
Index 171

Preface

Percutaneous absorption is the basis for the development of transdermal drug delivery patches, as well as treatment of skin disorders. The concept of a patch is now widely accepted in the general population, in part due to the heavy advertisement of nicotine patches as an aid for smoking cessation. The skin permeability of most drugs is low but can be enhanced by several mechanisms, including electric force. Iontophoresis utilizes a small amount of electric current to push the drug through the skin. Iontophoresis technology has been successfully used in clinical medicine to achieve topical delivery of drugs for several decades. The field of iontophoretic delivery is now gaining increasing recognition from the pharmaceutical industry for development of wearable patches for systemic delivery of drugs, with several drug companies already in phase II and phase III clinical trials. Wearable iontophoretic patches should be on the market within the next few years. Despite the challenges as discussed in this book, wearable iontophoretic patches are likely to be commonplace in the next century, with several of the drugs being therapeutic peptides. Only a handful of companies have the manufacturing expertise to produce electrically assisted transdermal patches owing to the large investment required. However, these companies are offering contract manufacturing to larger pharmaceutical companies so that the number of companies with iontophoretic or other electrically assisted patches on the market is expected to be large in the coming years. Microprocessors in these wearable patches may record the date and time of use, thus providing capabilities to monitor patient compliance as an additional advantage.

Another technique, electroporation, applies voltage pulses which open up new pores in the skin, and its applications can extend to larger molecules such as therapeutic proteins and oligonucleotides, the new products of biotechnology. The field of electroporation is also not new but its application to transdermal delivery is very new, and several companies are trying to commercialize this technology. Electroporation of skin is a relatively new area which still has many unanswered questions. At present, the technology does not allow the development of wearable patches. Perhaps, it may not find widespread use for common drug delivery applications but could be useful for treatment of debilitating illnesses such as cancer chemotherapy. It could also find applications in gene therapy of skin, and in immunization by DNA vaccines, as discussed in this book.

Over 750 references have been provided for the discussion in this book, including patent literature. While patent literature may not be peer reviewed, it can nevertheless provide some very useful and practical information. Some of the information was obtained via the Internet but independent verification through the company or other sources was obtained to verify the authenticity of the information to the best of the author's knowledge. The inevitable hype on some of the Internet sites was hopefully placed in perspective through my overall sense of the field from the literature and work in my laboratory. References to Internet address are not provided as such information changes very rapidly. Since the book was written as a single author book, I was hopefully able to prevent overlap of information between chapters. However, since related information may have been split between chapters, I have provided extensive cross-referencing information to the relevant section. As each chapter has its unique section numbering sequence, it should be easy to find this information.

The task undertaken in this work is wide in scope, and help was sought by sending selected chapters of this book to leading experts in the field for a review. The resulting feedback was incorporated into the text. I would like to acknowledge these individuals for their helpful suggestions to improve this book. Dr Philip Green of Becton Dickinson Transdermal Systems (Fair Lawn, NJ, USA) reviewed Chapters 2, 3, 5, and 8. Dr Green will be one of the key people responsible for bringing a wearable iontophoretic patch to market and I was fortunate to have him review almost half the book which was in his area of expertise. Dr Peter C. Panus of the Department of Physical Therapy, East Tennessee State University (TN, USA) reviewed Chapter 4 on applications of iontophoretic devices in physical therapy. Dr Panus has unique expertise in this area, having worked in this field both as a clinician and a researcher. Dr Mark R. Prausnitz of the School of Chemical Engineering, Georgia Institute of Technology (Atlanta, GA, USA) reviewed the topic of electroporation (Chapters 6 and 7), a field he was instrumental in pioneering. Thanks are also extended to Dr Prausnitz for helpful discussions over the last year or so about the experimental work in our laboratory with electroporation, which has not yet been published. My special gratitude also goes to Dr Srini Tenjarla of Mercer University Southern School of Pharmacy (Atlanta, GA, USA) who, in addition to providing a thorough critique of Chapter 1, read the whole book and provided constructive criticism. Dr Tenjarla is a well known researcher in the field of topical and transdermal delivery systems. Finally, I would like to acknowledge the staff of Taylor & Francis for their efforts in bringing this book to the readers.

Ajay K. Banga

About the author

Dr Ajay K. Banga is Associate Professor of pharmaceutics at the School of Pharmacy, Auburn University, AL, USA. He holds a Ph.D. degree from Rutgers – The State University of New Jersey, USA. Dr Banga has over 60 publications and presentations, mostly related to electrically assisted transdermal and topical delivery. He has also written another single author book, *Therapeutic Peptides and Proteins: Formulation, Processing, and Drug Delivery* (Technomic Publishing Company, 1996). He is a nationally and internationally invited speaker to companies and organizations and has also organized several seminars in Europe and the United States. Prior to becoming an academic, Dr Banga worked in the pharmaceutical industry at Ranbaxy Laboratories (N. Delhi, India) and Bausch & Lomb (Rochester, NY). Dr Banga has served as the chairman for the education committee of the BIOTEC section, and the past chair of the membership committee for the Pharmaceutics & Drug Delivery section of the American Association of Pharmaceutical Scientists.

1

Percutaneous absorption and its enhancement

1.1 Introduction

The skin is the largest organ of the human body, with a surface area of about 2 m^2. Historically, the skin was viewed as an impermeable barrier as its primary purpose is protection against the entry of foreign agents into the body. However, in recent years, it has been increasingly recognized that intact skin can be used as a port for topical or continuous systemic administration of drugs. Local/topical dermatological delivery is useful when the skin is the target site for the medication. On the other hand, skin can also be used as a route of administration for systemic delivery of the drug via a transdermal patch. For drugs which have short half-lives, a transdermal route provides a continuous mode of administration, somewhat similar to that provided by an intravenous infusion. However, unlike an intravenous infusion, delivery is non-invasive and no hospitalization is required. Once absorbed, the hepatic circulation is bypassed, thus avoiding another major site of potential degradation (Brown and Langer, 1988; Ranade, 1991). Currently, several drugs such as oestradiol, scopolamine, nitroglycerine, clonidine, fentanyl, nicotine and testosterone have been successfully marketed for transdermal delivery, with several brands available for some drugs. In addition, a salicylic acid patch is available for local treatment of plantar warts. In 1996, the global market for all transdermal products was estimated to be about US$ 1.5 billion. The commercialization of transdermal drug delivery has required technology from many disciplines beyond pharmaceutical sciences, such as polymer chemistry, adhesion sciences, mass transport, web film coating, and printing (Potts and Cleary, 1995). Electrotransport is another example where the science developed in other disciplines can be used to expand the scope of transdermal technology to include the delivery of macromolecules. For instance, while the use of electroporation to enhance penetration of drugs across skin or intact tissue is relatively new, the technique itself has been widely used to insert genes into cells for recombinant DNA work. Transdermal delivery cannot and need not be used in every situation. A rationale to explore this route exists only for drugs that are subject to an extensive first-pass metabolism when given orally or those that must be taken several times per day. Even then, only potent drugs can be administered through this route since there are economic and cosmetic reasons to not exceed the patch size beyond a certain limit. Though it is hard to make generalizations, the maximum patch size has been suggested to be about 50 cm² while the maximum possible dose that can be delivered may be around 50 mg per day (Guy, 1996).

1.2 Structure, enzymatic activity and immunology of the skin

1.2.1 Skin structure

The skin is composed of an outer epidermis, an inner dermis, and the underlying subdermal tissue. A basement membrane separates the epidermis and dermis, whereas the dermis remains continuous with the subcutaneous and adipose tissues (Berti and Lipsky, 1995). The dermis provides physiological support for the epidermis by supplying it with blood and lymphatic vessels and also with nerve endings. The epidermis comprises several physiologically active epidermal tissues and the physiologically inactive stratum corneum. The physiologically active epidermis contains keratinocytes as the predominant cell type. The skin also has other cell types which represent the non-keratinocytes. These include melanocytes (pigment formation), Merkel cells and Langerhans cells. Merkel cells reside just above the basement membrane and these scarce cells are assumed to mediate touch (Melski, 1996). Langerhans cells mediate the immunological function and are discussed in Section 1.2.3. The keratinocytes originate in a layer called the stratum germinativum and undergo continuous differentiation and mitotic activity during the course of migration upwards through the layers of spinosum, granulosum and lucidum. Finally, a layer of dead, flattened keratin-filled cells (corneocytes) which is called the stratum corneum or the horny layer, is produced. The stratum corneum is actually composed of about 10 to 15 layers of these flattened cornified cells. The entire epidermis is avascular and is supported by the underlying dermis. The cells migrating from the stratum germinativum layer are slowly dying as they move upwards away from their source of oxygen and nourishment. Upon reaching the stratum corneum, these cells are cornified and dead. The time required for the cells to proliferate from the stratum germinativum to the stratum corneum is about 28 days, of which 14 days' existence is as corneocytes in the stratum corneum layer. The corneocytes are then sloughed off from the skin into the environment (about one cell layer per day), a process called desquamation. There are several appendages in the skin, which include hair follicles and sebaceous and sweat glands, but these occupy only about 0.1 per cent of the total human skin surface (Banga and Chien, 1993b). The hair follicle, hair shaft and sebaceous gland form what is commonly termed the pilosebaceous unit. Infants have a higher pore density as the pores of the skin are fixed at birth and do not regenerate. Resident flora in pores is usually not composed of pathogens. The pilosebaceous follicles have about 10 to 20 per cent of the resident flora and cannot be decontaminated by scrubbing. Sebaceous glands secrete sebum, a mixture of triglycerides, phospholipids and waxes. The purpose of sebum is unknown. Sebaceous glands are absent on the palms, soles and nail beds. Sweat glands or eccrine glands respond to temperature via parasympathetic nerves, except on palms, soles and axillae, where they respond to emotional stimuli via sympathetic nerves (Melski, 1996). Electrokinetic data on skin suggests that there may be a wide distribution of pore sizes with different radius, charge, and charge concentration. Based on a model, the mean pore size for hairless mouse skin was estimated to be 16 Å (Pikal, 1995), while that for ethanol-pretreated human epidermal membrane was estimated to be 22–54 Å (Inamori et al., 1994). Using a microscopic model, the pore radius of human cadaver skin based on streaming potential and charge density data has been suggested to be about 200 Å (Aguilella *et al.*, 1994).

The total thickness of skin is about 2–3 mm, but the thickness of the stratum corneum is only about 10–15 μm. However, most of the epidermal mass is concentrated in the stratum corneum and this layer forms the principal barrier to the penetration of drugs. This rate-limiting barrier to transdermal permeation is packed with hexagonal cells, an arrangement which has provided a large surface area with the least mass. In contrast to most other epithelia, the barrier function of epidermis is not based on tight junctions between the cells but rather on the lipid lamellae of the stratum corneum. Species difference exists, for example the cells are stacked in vertical columns in mice but distributed randomly in humans. Each corneocyte is bounded by a thick, proteinaceous envelope with the tough fibrous protein keratin as the main component. Earlier reports based on transmission electron microscopy suggested that the spaces between corneocytes are empty; however, this is now believed to be an artifact of sample preparation. In fact, an intact stratum corneum is a highly ordered structure. Recent freeze-fracture electron microscopy studies on hydrated human stratum corneum have confirmed that corneocytes are embedded in intercellular lipids, and are aligned parallel to the surface of the stratum corneum (van Hal *et al.*, 1996). As the epithelial cells migrate upwards towards the stratum corneum, their plasma membrane seems to thicken owing to a deposition of material on their inner and outer surfaces. This is the process of keratinization during which the polypeptide chains unfold and break down and then resynthesize into keratin, the tough fibrous protein which forms the main component of the corneocyte. With increasing understanding of the skin, it is now being recognized that keratin represents a group of proteins which play a very important role as scaffolding filaments within epithelial cells (Morley, 1997). Corneocytes contain a compact arrangement of α-keratin filaments 60–80 Å in diameter and distributed in an amorphous matrix. The intercellular spaces of the stratum corneum are completely filled with broad, multiple lamellae. The lipids constituting these lamellae are composed of ceramides, cholesterol, and free fatty acids in approximately equal quantities (Fartasch, 1996). In contrast, the lipid composition in the viable epidermis is predominantly phospholipid. These changes in the lipid composition have been demonstrated to occur during the keratinization process of the stratum corneum. The morphology of these lipids plays an important role in the barrier function of the stratum corneum (Denda *et al.*, 1994). The dermis is composed of a loose connective tissue and contains collagen and elastin fibres. In the dermis, glycosaminoglycans, or acid mucopolysaccharides, are covalently linked to peptide chains to form proteoglycans, the ground substance that promotes the plasticity of the skin. In the ground substance, spindle-shaped cells called fibroblasts are interspersed between collagen bundles. In addition, mast cells are present in the ground substance (Melski, 1996). Nerves, blood vessels and lymphatic vessels are also present in the dermis. Epidermal appendages such as hair follicles and sweat glands are embedded in the dermis (Steinstrasser and Merkle, 1995).

1.2.2 Enzymatic activity of the skin

The skin has considerable enzymatic activity, including cytochrome P450 isozymes which may be localized to specific cell types, especially in the epidermis and pilosebaceous system. Enzymes identified in the stratum corneum include those with lipase, protease, phosphatase, sulphatase and glycosidase activity (Howes *et al.*, 1996). Xenobiotic substances are first chemically activated by oxidation with the involvement of cytochrome P450 isozymes. These enzymes are localized mainly in the endoplasmic reticulum and the activity is highest in the microsomal fraction of skin homogenates. The catalytic activity of the enzymes in hair follicles is particularly high. While epidermal activities of cytochrome P450 in skin are only about 1–5 per cent of those in liver, the transferase activity in skin can be as high as 10 per cent of hepatic values (Merk *et al.*, 1996). The total skin blood flow is only about 6.25 per cent of the total liver blood flow. Thus, the metabolism is lower in skin though the spectrum of reactions in the skin is similar to those observed in the liver (Tauber, 1989). The enzymatic activity of the skin varies with the anatomical site. For instance, hydrocortisone 5α-reductase activity was detected only in the human foreskin while high levels of testosterone 5α-reductase were found in the scrotal skin. The distribution of enzymatic activity within the various skin layers is not well known owing to difficulties with the experimental methodologies. As blood capillaries lie just under the epidermis–dermis junction, drugs may only have minimal contact with dermal enzymes before they are taken up by the general circulation. Thus, the enzymatic activity of the epidermis may be more important as a barrier to drug absorption (Steinstrasser and Merkle, 1995). The enzymatic hydrolysis of drugs in the skin has been reported (Valia *et al.*, 1985) and may differ between *in vivo* and *in vitro* conditions, with *in vitro* results sometimes overestimating metabolism because of increased enzymatic activity and/or lack of removal by capillaries (Guzek *et al.*, 1989; Potts *et al.*, 1989). The proteolytic activity of the skin is discussed separately in Section 5.4.1.

1.2.3 Immunology of the skin

The skin performs a complex defence function which may be described as 'immunological'. The capacity of the skin to distinguish self from non-self is remarkable if one considers the vast variety of exogenous substances to which the skin is continuously exposed. Our understanding of skin immunology has now advanced tremendously, as illustrated by the publication of a recent text (Bos, 1997b). The immunological environment of the skin including the humoral and cellular components is given the acronym SIS (skin immune system). Dysregulations of this system can manifest themselves as immunodermatological diseases, including atopic eczema, psoriasis, cutaneous lupus erythematosus, scleroderma, and autoimmune bullous disease (Zierhut *et al.*, 1996; Bos, 1997a). It is known that Langerhans cells reside in the epidermis and express a high level of major histocompatibility complex (MHC) class II molecules and strong stimulatory functions for the activation of T lymphocytes. The Langerhans cells comprise 2–4 per cent of the cells of the epidermis and are also found in lymph nodes. They act on antigens and present them to lymphocytes and thus provide immune surveillance for viruses, neoplasms and non-autologous grafts. The keratinocytes also play a role in immunity (Melski, 1996). The role of Langerhans cells in the DNA vaccination of skin by electroporation is discussed in Section 7.4. The Langerhans cells are dendritic-shaped cells which are located in the basal parts of the epidermis. In recent years, the concept of skin-associated lymphoid tissue or 'SALT' has evolved in which Langerhans cells in the epidermis are believed to act as antigenic traps, and the antigen-laden cells then migrate into dermal lymphatic channels to present the information to T lymphocytes in lymph nodes. When allergens penetrate into the skin, they can in some cases lead to allergic contact dermatitis, which is characterized by redness and vesicles, followed by scaling and dry skin. Additional relevant compounds in skin immunology are the eicosanoids. Eicosanoids, which are oxygenated metabolites of 20-carbon fatty acids, especially arachidonic acid, are a class of compounds which have a role in the pathophysiology of inflammatory and immunological skin disorders. For example, leukotrienes play a central role in the pathogenesis of psoriasis, a chronic, scaly and inflammatory skin disorder (Bos, 1997b). The immunology of the skin needs to be carefully considered to design iontophoretic systems. Some aspects which need to be considered are discussed in Chapter 8.

1.3 Percutaneous absorption

1.3.1 Mechanisms of percutaneous absorption

The stratum corneum is a predominantly lipophilic barrier which minimizes transepidermal water loss. It is also the principal permeation barrier to percutaneous absorption. The transdermal delivery of small molecules has thus been considered as a process of interfacial partitioning and molecular diffusion through this barrier. A typical mathematical model treats the stratum corneum as a two-phase protein–lipid heterogeneous membrane having the lipid matrix as the continuous phase (Michaels *et al.*, 1975; Elias, 1983, 1988). Several theoretical skin-permeation models have been proposed which predict the transdermal flux of a drug based on a few physicochemical properties of the drug (Tojo, 1987; Potts and Guy, 1992; Pugh and Hadgraft, 1994; Kirchner *et al.*, 1997; Lee *et al.*, 1997). These models often make some assumptions about the barrier properties of the skin and predict the transdermal flux of a drug from a saturated aqueous solution, given a knowledge of the water solubility and molecular weight of the drug and its lipid–protein partition coefficient. Most of these theoretical expressions assume a two-compartment model of the stratum corneum based on a heterogeneous two-compartment system of protein-enriched cells embedded in lipid-laden intercellular domains. An analogy of 'bricks and mortar' is often given for this model. Based on this model, drugs can diffuse through the stratum corneum via a transepidermal or a transappendageal route. Transepidermal drug penetration through the stratum corneum can take place between the cells (intercellular) or through the protein-filled cells (transcellular route). The relative contribution of these routes depends on the solubility, partition coefficient and diffusivity of the drug within these protein or lipid phases. The transappendageal route normally contributes only to a very limited extent to the overall kinetic profile of transdermal drug delivery. Recently, it was shown that the penetration of retinoic acid was greater through the skin of hairless guinea pigs than those with hair, confirming that the structure and composition of the stratum corneum is more important than follicular density for passive permeation (Hisoire and Bucks, 1997). However, the hair follicles and sweat ducts can act as diffusion shunts for ionic molecules during iontophoretic transport (Figure 1.1). A detailed discussion of the mechanisms of iontophoretic delivery can be found in Chapter 2. The pathways of electroporation and electroincorporation as seen in Figure 1.1 are discussed in Chapter 6. There has been some recent interest in transfollicular drug delivery because the origin of several dermatological problems such as acne and alopecia relate to the hair follicle (Lauer *et al.*, 1995).

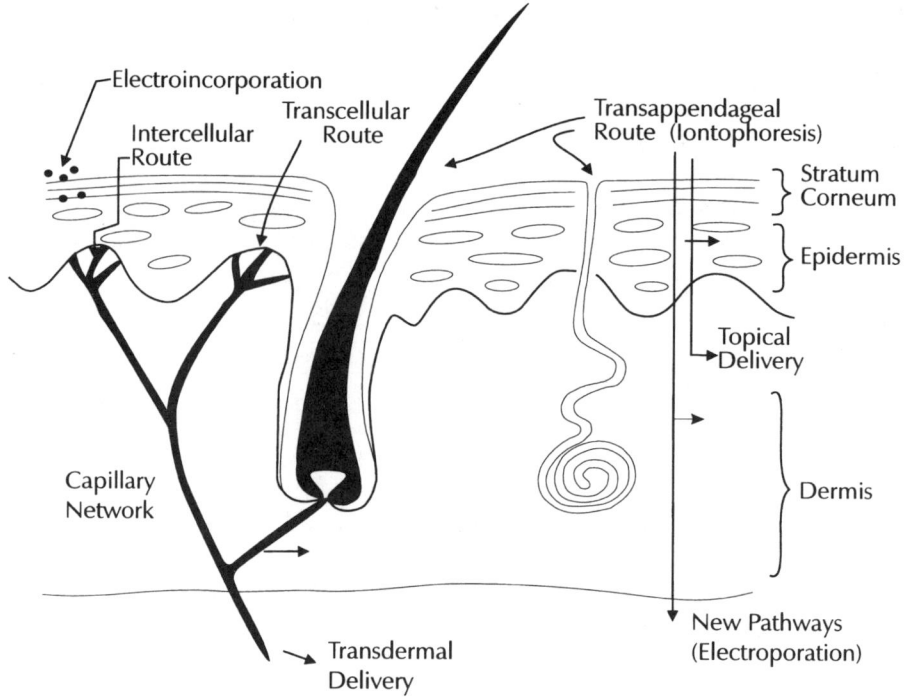

Figure 1.1 A schematic showing the pathways of topical and transdermal delivery, including electrically assisted delivery by iontophoresis, electroporation or electroincorporation.

1.3.2 Theoretical basis of percutaneous absorption

The passive diffusion of a non-electrolyte in the absence of any bulk flow is expressed by Fick's first law of diffusion as:

$$J = -D \, dC/dx \tag{1.1}$$

where J is the flux, D is the diffusion coefficient and dC/dx is the concentration gradient over a distance x. Fick's first law of diffusion can be used to describe the skin permeation of drugs; however, the concentration gradient across the skin tissue cannot be easily measured but can be approximated by the product of the permeability coefficient (P_s) and the concentration difference across the skin (C_s). The steady-state transdermal flux, J_s, through skin barrier is thus given as:

$$J_s = P_s \, C_s \tag{1.2}$$

where P_s, the permeability coefficient, is defined by:

$$P_s = K.D/h \tag{1.3}$$

where K is the partition coefficient and h is the thickness of the skin. The cumulative amount of drug permeating through the skin (Q_t) is given by:

$$Q_t = K.D.C_s/h \, (t - h^2/6D) \tag{1.4}$$

where C_s is the saturated reservoir concentration when a sink condition is maintained in the receptor solution. Differentiation of Eqn 1.4 with respect to time will yield Eqn 1.2, which describes the steady-state transdermal flux. When the steady-state line is extrapolated to the time axis, the value of lag time, t_L, is obtained by the intercept at $Q = 0$.

$$t_L = h^2/6D \qquad (1.5)$$

The intercept, t_L, is a measure of the time it takes for the penetrant to achieve a constant concentration gradient across the skin. The lag time method is commonly used for analysis of permeation data from *in vitro* experiments with an infinite dosing technique, that is, where the skin separates an infinite reservoir of drug on the donor side and a perfect sink as receptor. The steady-state flux is calculated from the slope of the linear permeation profile, while the *x*-intercept provides the lag time of diffusion. The diffusion coefficient can then be calculated from Eqn 1.4, by knowing the donor phase concentration and thickness of the barrier, and measuring the partition coefficient. However, the lag time method could be subjective as it requires judgement to determine the linearity of the permeation profile (Shah, 1993). As the diffusion coefficients of commonly used drugs across the skin range from 10^{-6} to 10^{-13} cm^2/s, lag times across the skin can range from a few minutes to several days. Since the structural integrity of human cadaver skin for *in vitro* studies over such prolonged periods is questionable, it would be desirable to have low lag times. More importantly, low lag times are required for most therapeutic indications. Electrical enhancement of percutaneous penetration by iontophoresis and electroporation can be one such means to lower lag times. The theoretical basis of iontophoresis will be discussed in Chapter 2, though the topic is being introduced in this chapter.

1.4 Iontophoretic enhancement of percutaneous absorption

Transdermal and topical delivery of drugs, that is, delivery of drug through or into the skin, can be assisted by electrical energy. The physical force involved can be iontophoresis, electro-osmosis or electroporation (Riviere and Heit, 1997). Electro-osmosis is a phenomenon that accompanies iontophoresis and will be discussed as we start discussing iontophoresis in detail starting from Chapter 2. Electroporation, on the other hand, is a different mechanism and will be discussed in Chapters 6 and 7.

1.4.1 *Iontophoresis*

Iontophoresis implies the use of small amounts of physiologically acceptable electric current to drive ionic (charged) drugs into the body (Singh and Roberts, 1989; Green *et al.*, 1993; Singh and Maibach, 1994a, b; Singh and Bhatia, 1996). By using an electrode of the same polarity as the charge on the drug, the drug is driven into the skin by electrostatic repulsion. The technique is not new and has been used clinically in delivering medication to surface tissues for several decades. However, its potential is recently being rediscovered for transdermal systemic delivery of ionic drugs including peptides and oligonucleotides which are normally difficult to administer except by the parenteral route. The technique has been observed to enhance the transdermal permeation of ionic drugs severalfold, and this can expand the horizon of transdermal controlled drug delivery for systemic medication. Besides the usual benefits of transdermal delivery, iontophoresis presents a unique opportunity to provide programmed drug delivery. This is because the drug is delivered in proportion to the current, which can be readily adjusted. Such dependence on current may also make drug absorption via iontophoresis less dependent on biological variables, unlike most other drug delivery systems. Also, patient compliance will improve and, as the dosage form includes electronics, means to remind patients to replace the dose can be built into the system. The dose can also be titrated for individual patients by adjusting the current. Several new developments have taken place in the field of iontophoretic drug delivery since the topic was last reviewed by the author 10 years ago (Banga and Chien, 1988). A wearable iontophoretic electric patch is now closer to being marketed, partly due to advances in microelectronics. The technology is presently in clinical development and miniature, battery powered and wearable patches are predicted to be on the market prior to the year 2000 (Green, 1996b). Before the commercialization of microelectronics, workers had struggled for several decades with vacuum-tube control systems, lead-acid battery and relay-switching technology and thus wearable patches were not feasible. However, now the revolution in microelectronics makes the technology feasible and patch designs capable of 24-h use have been developed (Sage, 1993). Iontophoretic devices for systemic delivery are currently in development and will consist of a dosage unit the size of a traditional 'transdermal patch', with miniaturized circuitry and button cells. Larger devices may be shaped like a wristwatch and have programmable features to adjust dosage. Smaller prototypes have been developed and may be of the disposable or reusable type. In reusable systems, the drug may be contained in a hydrogel pad, which can be replaced as required. For disposable systems, perhaps the microprocessor can be removed and transferred to another patch, to keep costs low. Companies such as the Alza Corporation (Palo Alto, CA, USA), Becton Dickinson (NJ, USA) and others (see Chapter 8) are currently developing these patches and investigations are in an advanced stage of clinical trials. Iontophoresis, of course, constitutes the major topic of this book and the reader is referred to subsequent chapters for details. A brief overview of the historical origin of iontophoresis will be discussed in this chapter.

1.4.2 *Historical origin of iontophoresis*

When the Greek physician, Etius, prescribed the shocks of torpedo, an electric fish, for the treatment of gout, he set the stage for the biomedical applications of electricity even before the discovery of electricity! The idea of applying electric current to increase the penetration of electrically charged drugs into surface tissues probably originated from Veratti in 1747. A series of experiments were conducted over the next two centuries which were rather dramatic but would not be considered scientific or ethical today owing to more awareness of the humane usage of animals or human subjects. These are briefly discussed here as they have been reviewed by the author earlier (Chien and Banga, 1989a). In the latter part of the 19th century, Morton was interested in the electrical transport of drugs through the skin and wrote a book in 1898 on cataphoresis of ions into the tissues. He conducted an experiment on himself in which finely powdered graphite was driven into his arm using a positive electrode, producing small black spots which persisted for several weeks. The first well-documented experiments were done at the beginning of the 20th century by Leduc. He reproduced an experiment performed earlier by placing two rabbits in series with a direct-current generator, and demonstrating the introduction of strychnine and cyanide ions into the rabbits when the correct polarity was applied. The results were rather dramatic: the first rabbit was seized by tetanic convulsions, due to the introduction of strychnine ions, while the second rabbit died with symptoms of cyanide poisoning. In another experiment done by Inchley in 1921, a zinc–silver couple was applied to an anaesthetized cat's tongue in which the lint sandwiched between the tongue and the zinc plate was presoaked in a solution of atropine sulphate. Complete paralysis of the cranial autonomic nerves occurred in 5 to 10 min. Perhaps the first therapeutic use of iontophoresis came in the form of good news for patients with sweaty palms in 1936, when Ichihashi noted that sweating could be reduced by ion transfer of certain applied solutions by electrophoretic techniques. This was the origin of the application of iontophoresis for the treatment of hyperhidrosis, a condition characterized by excessive sweating, which can be socially and occupationally distressing. The use of iontophoresis for the treatment of hyperhidrosis is discussed in Section 4.2.5. While investigators were busy trying to reduce sweating, Gibson and Cooke had an altogether different problem. They wanted to increase sweating in order to obtain enough sweat for the diagnosis of cystic fibrosis, as it was known that cystic fibrosis patients have a high concentration of sodium and chloride in their sweat. To avoid a painful intradermal injection of a cholinergic drug, the simplest method available to increase sweating was to place the patient in a plastic bag. The procedure was obviously uncomfortable and required a long period of time. Besides, some infants became hyperpyrexic within 30 min. Alternatively, Gibson and Cooke used iontophoretic application of pilocarpine to induce sweating. The procedure was found painless and required only 5 min; rapid sweating was induced and continued for 30 min (Gibson and Cooke, 1959). Copper electrodes were suggested for the procedure, while the use of steel electrodes was associated with burns occurring during the test (Gibson, 1967). Following the discovery of the use of iontophoresis to induce sweating, additional studies were conducted and iontophoresis of pilocarpine has been approved by the United States Food and Drugs Administration (FDA) for diagnosis of cystic fibrosis (Section 4.2.6).

1.4.3 Other applications of iontophoresis

It should be realized that iontophoresis is useful for more than just skin (topical dermatological or transdermal) applications. Though only skin applications are discussed in this book, other applications are briefly introduced here. Ocular iontophoresis has been used for a long time (Erlanger, 1954) and widely investigated for the delivery of drugs such as gentamicin (Barza et al., 1987), cefazolin (Barza et al., 1986), fluorescein (Maurice, 1986), tobramycin, lidocaine, adrenalin, timolol maleate, idoxuridine, dexamethasone and others for several potential clinical applications (Hill et al., 1993; Sarraf and Lee, 1994). Iontophoresis of fluoride has been used for tooth densensitization (Brough et al., 1985; Gangarosa, 1986). Also, microiontophoresis is widely used in neuropharmacological and basic studies with neurons (Orsini et al., 1985; Skydsgaard and Hounsgaard, 1996; Warren and Dykes, 1996; Roychowdhury and Heinricher, 1997). Microiontophoresis involves the controlled ejection of drugs from micropipettes (tip diameter about 1 µm), allowing the study of effects and interactions of drugs on a very restricted area of tissue (Stone, 1972). Microiontophoresis can thus be used to administer neuroactive compounds such as neurotransmitters to single neurons to observe their effect on firing parameters. This allows the observation of effects on single neurons without affecting the whole nervous system, such as when the drug is given via systemic administration. It can also be used to deposit dyes or tracers into the cytoplasm of a cell or into the intercellular space for subsequent histological examination. The development and improvements in microiontophoretic electrodes (Fu and Lorden, 1996) are beyond the scope of this book. Iontophoresis can also find applications in the modulation of drug delivery from medical devices which may be implanted or introduced into the body. The regional nature of several cardiac diseases can benefit from localized drug delivery from an implant, which may be modulated by iontophoresis (Labhasetwar and Levy, 1995). Iontophoretic cardiac implants have been successfully used in dogs for delivery of the antiarrhythmic agent sotalol. This can allow for timely and localized delivery to the heart, avoiding problems of drug toxicity and bioavailability limitations (Avitall, 1992; Labhasetwar *et al.*, 1995; Schwendeman *et al.*, 1995). Suitable membranes to modulate iontophoretic release from such implantable systems are under development (Schwendeman *et al.*, 1994). Iontophoresis also has potential applications in wound healing as electric fields have been reported to enhance skin wound repair (Cheng *et al.*, 1996a; Goldman and Pollack, 1996).

1.5 Other methods for enhancement of percutaneous absorption

The skin is impermeable to macromolecules such as peptides, proteins, oligonucleotides and DNA. The physical and enzymatic barriers, in combination, create a formidable barrier for any permeation under normal circumstances. Various physical, chemical and biochemical techniques have been attempted to surmount these barriers. The iontophoresis technique briefly discussed above and other electrical enhancement techniques such as electro-osmosis and electroporation form the subject matter of this book and will be discussed in detail in subsequent chapters. In the following sections, some of the other techniques are discussed very briefly to provide the reader with an awareness of other methods for enhancement of percutaneous absorption. Invasive techniques such as subcutaneous injection and implants are not discussed here. Also, chemical modification of the drug to improve its percutaneous absorption is not discussed here. A novel technique which uses a transdermal delivery system with jet injection has also been reported but may also be considered invasive as a jet injector containing physiological saline first makes a pore in the skin prior to application of the drug (Inoue *et al.*, 1996). Alternatively, the drug can also be directly delivered to the skin in dry powder form using a supersonic flow of helium gas to accelerate the particles to a high velocity (Muddle *et al.*, 1997).

1.5.1 Phonophoresis

Phonophoresis implies the transport of drug molecules under the influence of ultrasound (Tyle and Agrawala, 1989). Though widely used by physiotherapists, systematic studies of its therapeutic value are needed (Meidan *et al.*, 1995). However, it has been suggested that low frequency ultrasound can even deliver therapeutic doses of proteins such as insulin, interferon and erythropoietin (Mitragotri *et al.*, 1995a). The drug is delivered from a coupling (contact) agent which transfers ultrasonic energy from the ultrasonic device to the skin. The exact mechanism involved is not known, but enhancement presumably results from thermal, mechanical and chemical alterations in the skin induced by ultrasonic waves. Confocal microscopy results indicate that cavitation occurs in the keratinocytes of the stratum corneum upon ultrasound exposure. The oscillations of the cavitation bubbles possibly enhance transdermal transport by inducing disorder in the stratum corneum lipids (Mitragotri *et al.*, 1995b). The proposed mechanisms for *in vitro* and *in vivo* phonophoresis have been reviewed in the literature (Simonin, 1995). Ultrasonic irradiation of the skin will change the electrochemical properties of the skin. It has been reported that constant-current iontophoresis (0.1 mA/cm^2) after pretreatment of excised hairless rat skin by ultrasound significantly increased the flux of benzoate anion (BA)

Percutaneous absorption and its enhancement 11

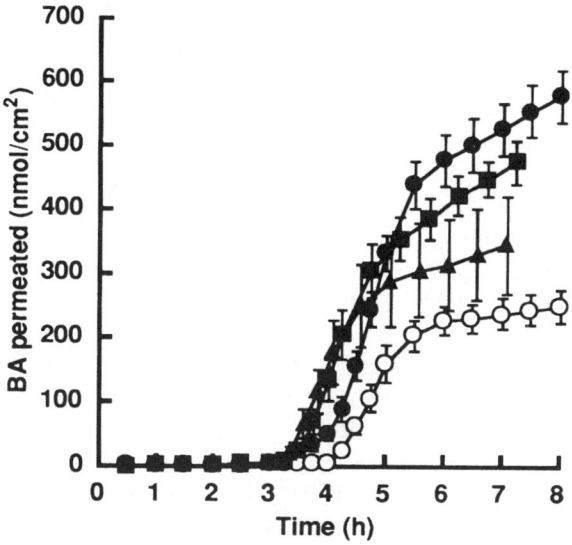

Figure 1.2 Effect of iontophoresis on the ultrasonic pretreated skin. The symbols represent iontophoresis with ultrasonic pretreatment for 0 (○), 5 (▲), 15 (■), and 60 min (●). Each data point represents the mean ± S.E. of three experiments. (Reprinted from *International Journal of Pharmaceutics*, **137** Ueda *et al.*, Change in the electrochemical properties of skin and the lipid packing in stratum corneum by ultrasonic irradiation, pp. 217–224, Copyright 1996 with kind permission from Elsevier Science – NL, Sara Burgerhartstraat 25, 1055 KV Amsterdam, The Netherlands.)

through the skin compared with that without pretreatment. The flux during iontophoresis was dependent on the duration of ultrasonic pretreatment (Figure 1.2). The difference in electrical potential across the skin during iontophoresis was lower if the skin was pretreated with ultrasound, suggesting that ultrasound causes structural disorder in the stratum corneum lipids, leading to an increased aqueous region in the stratum corneum and increased diffusivity of benzoate ion in the skin (Ueda *et al.*, 1996).

1.5.2 *Chemical penetration enhancers*

The use of chemical penetration enhancers to improve percutaneous absorption of drugs has been reviewed by the author (Ghosh and Banga, 1993a, b) and more recent reviews are also available (Walker and Smith, 1996). A penetration enhancer is usually a small molecule and mechanisms of enhancement include modification of the intercellular lipid matrix with increased membrane fluidity. Examples of commonly investigated chemical enhancers include bile salts, chelating agents, surfactants, fatty acid derivatives, alkanols, alkanoic acids and their esters, dimethylsulphoxide (DMSO), 1-dodecylazacycloheptan-2-one (azone) and cyclodextrins. While penetration enhancers may be promising for some smaller molecules, their use is unlikely to be successful for delivery of macromolecular drugs. Also, as enhancers modify the structure of the skin (Barry, 1987, 1991), reversibility of effect and safety of long-term usage will have to be demonstrated to obtain regulatory approval. In addition to increasing the absorption of the drug, enhancers may non-specifically increase their own absorption or that of formulation excipients. An ideal enhancer must be non-toxic, non-irritating, non-allergenic, pharmacologically inert and compatible with most drugs and excipients. Currently, azone and oxazolidinone are considered to be among the most promising enhancers. However, no single agent meets all the desirable attributes of an enhancer. A combination of enhancers may thus be required. In addition to these chemical enhancers, many of the generally recognized as safe (GRAS) parenteral vehicles can also enhance percutaneous drug absorption. Many of the transdermal and dermal delivery products on the market are formulated with cosolvents (Pfister and Hsieh, 1990b). These cosolvents include propylene glycol, polyethylene glycol 400, isopropyl myristate, isopropyl palmitate, ethanol, water and mineral oil. Since an enhancer is delivered to the skin, its pharmacokinetics must be determined so as to know its half-life in skin, degree of absorption, mechanism of elimination, and metabolism. Also, the reversibility of skin barrier properties should be determined as any breach of the barrier properties of the stratum corneum could result in infection (Pfister and Hsieh, 1990a). Since use of a transdermal patch creates occlusive conditions on the skin, this leads to increased hydration and irritation of skin underneath the patch. As discussed, increased skin hydration may increase the permeation of the enhancer itself, which may cause even more irritation or toxicity. Further, the enhancer may increase the permeation of the formulation components along with the drug. Therefore, a careful evaluation of the long-term local and systemic toxicity of the chemical enhancers in the final transdermal dosage form is required (Ghosh and Banga, 1993a).

1.5.3 Combined use of enhancers and iontophoresis

Chemical and electrical enhancement are the two major techniques to facilitate transdermal drug delivery and their combined use may offer some additional advantages (Ganga et al., 1996). A combination of several penetration enhancers with iontophoresis has been investigated for the transdermal delivery of nonivamide acetate through rat skin. Cetylpyridinium chloride and isopropyl myristate were the most effective, but any potential clinical use of the latter will be limited as it induced severe changes in the histological structure of the skin (Fang et al., 1997). In another in vitro study on transdermal delivery of sotalol, enhancers were as effective as iontophoresis. A combination of enhancers and iontophoresis did not further increase permeation but actually decreased it slightly compared with enhancers or iontophoresis alone (Hirvonen et al., 1993). The effect of ethanol on the in vitro iontophoretic transport of a dopamine agonist has also been investigated. It was observed that ethanol increased the solubility of the agonist in water by as much as fivefold, but this only results in a modest increase in the amounts transported across the skin into the receptor compartment. A significant amount of the agonist was found to be retained in the skin (Hager et al., 1993). Any potential interaction between chemical and electrical enhancement should be investigated by techniques such as impedance spectroscopy (Kalia and Guy, 1997) or X-ray diffraction (Chesnoy et al., 1996). The combined use of enhancers and iontophoresis for delivery of peptides is discussed separately in Section 5.4.3.

2

Iontophoretic transdermal drug delivery

2.1 Introduction

An introduction to iontophoresis and its historical background was presented in Chapter 1. In this chapter, details on iontophoretic delivery such as pathways, mechanism, theory, factors affecting delivery, and some case studies will be presented. It is important to understand some of the very basic terminology used in the field. The electric current from a power source is delivered to the solution containing the drug where it is converted to an ionic flow taking place through the solutions and skin. The terminals leading the current into and out of the solution are electrodes, the positive pole being the anode and the negative pole being the cathode. The electrode itself may be smaller than the drug reservoir that is in contact with the skin. The current density is defined as the current intensity per unit cross-sectional area. In general, the current density will vary from point to point in any system, and the value calculated from the amperes divided by the surface area is just an average value at the treatment surface.

2.2 Electrical properties of skin

The electrical properties of the epidermal stratum corneum were investigated relatively early by tape stripping, when it was observed that removal of the stratum corneum dramatically reduced the observed resistance (Yamamoto and Yamamoto, 1976). This suggests that the stratum corneum forms the high electrical resistance layer and is a very important element for the skin impedance. The high resistance of this layer in turn is owing in part to its lower water content (about 20 per cent) compared with the normal physiological level (about 70 per cent). To understand the concept of impedance in rather simplistic terms, we need to first understand the term 'capacitance'. An arrangement of parallel plates separated by a very small distance and connected to a battery allows electrons to distribute over the lower plate. The electrons on the lower plate then induce a positive charge on the upper plate so that more electrons can now flow into the lower plate from the battery. This arrangement of parallel plates acts as a capacitor or condenser and this property is called capacitance. Thus, the capacitance of a capacitor is its ability to store an electric charge, flowing into it in the form of current. Biological tissues, such as skin tissue, also have a capacitance because of their ability to store electric charge and are thus electrical capacitors. When an electric circuit contains both capacitive and resistant elements, it is said to be reactive, in contrast to the one containing resistant elements only, which is said to be resistive. The equivalent circuit model for the stratum corneum employs a parallel arrangement of a resistor and a capacitor or a resistor in series with a parallel combination of a resistor and a capacitor (Yamamoto and Yamamoto, 1976). A reactive circuit is said to present impedance rather than resistance. The impedance represents the total electrical opposition of the circuit to the passage of a current through it. Measurements of electrical impedance of the skin are important in understanding electrically assisted transdermal delivery of drugs and also for *in vivo* electrical measurements on the body such as EEG and ECG (Emtestam and Ollmar, 1993; Pliquett and Pliquett, 1996). The human skin reportedly shows a high impedance to alternating current of low frequency but this decreases as frequency is increased (Plutchik and Hirsch, 1963). Skin resistance also decreases with the application of square voltage or current pulses in the range of 1 to 10 mA, which are frequently used for transcutaneous electrical stimulation of nerves. Owing to variations in skin resistance, the current intensity of voltage-regulated stimuli cannot be easily controlled so that only current-regulated stimuli are suggested for use as the current through the tissue is the most significant stimulus parameter (Boxtel, 1977). The skin and the electrode, together with any gel or contact material used, constitute an interface that has an electrode–skin impedance which should be considered when measurements are made (Calderwood, 1996). The loss of skin resistance with application of current may be due to a reorientation of molecules along the ion transport pathways, such as the possible realignment of lipid molecules in hair follicles and sweat glands (Kalia and Guy, 1995). Also, application of current will lead to an increase in the local ion concentration which will result in reduced resistance. The impedance of excised nude mouse skin, at 0.2 Hz, decreased by a factor of about 5 when exposed to iontophoretic current (0.16 mA/cm^2 for 1 h) during a period of hydration (8 h) compared with skin which underwent only hydration (Burnette and Bagniefski, 1988). The electrical resistance of excised human skin was measured at 0.2 Hz and was found to decrease by an order of magnitude when a current of 0.16 mA/cm^2 was applied for 1 h. The resistance then recovered but the plateau value was lower than the resistance of the skin before the current was applied (Burnette and Ongpipattanakul, 1988).

The current–voltage relationship in skin has been known to be non-linear for a long time (Stephens, 1963) and has been more thoroughly investigated in recent years for excised skin (Kasting and Bowman, 1990a, b). In a study with excised, full-thickness, hairless rat skin, those samples subjected to hypotonic solutions were most conductive and were characterized by highly symmetrical current–voltage profiles, while those subjected to isotonic or hypertonic conditions were less conductive and characterized by highly asymmetric current–voltage profiles (Ruddy and Hadzija, 1995). Temperature also affects the impedance of the skin. Using excised hairless mouse skin, it was shown that with increasing temperature, resistance decreased while capacitance increased, with most significant changes at the phase transition temperature (60°C) of the stratum corneum lipids. The impedance became independent of frequency, suggesting that the capacitive properties of the barrier were lost (Oh *et al.*, 1993). The modern bioengineering tools to measure electrical properties of skin and the effect of iontophoresis on such properties are discussed in Section 8.5.2.

2.3 Pathways of iontophoretic drug delivery

It has been known for a long time that sweat glands play a role in transport during iontophoresis. An early study showed that pore patterns developed on the skin following iontophoretic transfer of basic and acidic dyes and metallic ions (Abramson and Gorin, 1940). For example, thorough rubbing and washing of the skin following the iontophoretic delivery of methylene blue revealed a remarkable pattern of channels traversed by the dye. The blue dots observed on the skin were found to be the sites of the pores of the skin which are the orifices of the coils of sweat glands, suggesting that the dye enters the skin via these pores. The pore patterns persisted for several weeks in many cases. Similarly, fluorescein dye has been shown to penetrate excised human skin upon applying a current density of 0.16 mA/cm^2 and appeared on the dermal surface as spots at pore sites (Burnette and Ongpipattanakul, 1988). A comprehensive review of the pathways of iontophoretic current flow through mammalian skin has been published (Cullander, 1992). The macropores and other conductive pathways in skin have been studied by scanning electrochemical microscopy during iontophoresis, and by theoretical considerations. The current maximum was found near the exit of a hair follicle and the micropore radius determined was 9.2–14.1 µm. The macropore was considered to be a long cylindrical tube which was closed at one end. As the current was applied, the charging of the capacitance of one or two layers of macropore walls was the driving force for electroactivation. The free energy of the system was reduced, which pulled water into the tube to open it gradually. The opening time was 30 min for a 4-mm long macropore (Kuzmin *et al.*, 1996; Melikov and Ershler, 1996). During iontophoresis, the greatest concentration of ionized species is expected to move into some regions of the skin where there is damage, or along the sweat glands and hair follicles, as the diffusional resistance of the skin to permeation is lowest in these regions. Thus, a pore pathway is generally assumed for iontophoretic delivery. Iontophoresis of pilocarpine is used to induce sweating in the diagnosis of cystic fibrosis (Section 4.2.6), suggesting that some drug probably travels down the eccrine duct. In a study with desglycinamide arginine vasopressin (DGAVP) as a model peptide, its transport across human stratum corneum and snake skin was compared to assess the role of appendages such as hair follicles and sweat and sebaceous glands, which are present in human skin but absent in shed snake skin. While the initial resistance of both human and snake skin were in the same order of magnitude (about 25 kΩ–cm^2), the steady-state iontophoretic DGAVP flux across human stratum corneum was about 140 times larger than through shed snake skin. Also, the average lag time across human stratum corneum was 0.7 h, while that across shed snake skin was 2.5 h. Azone pretreatment of the skin led to a large increase in transport across snake skin but not human skin. This suggests that the intercellular lipid pathway contributes very little to the iontophoretic flux across human skin but is very important for snake skin (Hinsberg *et al.*, 1995). Using special electrodes, it has been suggested that the dominant pathway for flow of electric current through skin is through the sweat ducts. This study used a very thin (0.15 mm diameter) wire which was fine enough to distinguish between most pores and a very thin (0.1 µm) metal film electrode. The film electrode was placed on the skin and was permanently marked by the pathways of current flow so that dots developed after some seconds at places with sweat duct units (Grimnes, 1984). The 'aqueous pathway' for iontophoretic delivery has been reinforced by a study which observed the transport kinetics of an anion (salicylate), a cation (phenylethylamine), a polar neutral compound of low molecular weight (mannitol) and a polar neutral compound of high molecular weight (inulin). Using both intact and stripped dermatomed excised human skin, iontophoresis enhanced the delivery of all compounds relative to passive transport and the skin was shown to be both ion and size selective (Singh et al., 1995).

It should be noted that the pore pathway for delivery does not necessarily imply skin appendages only. Using a scanning electrochemical microscope, reddish-brown spots have been visualized in the skin for transdermal flow of iron. These spots may represent precipitation or complexation of iron within a localized pore-type region but were not associated with a skin appendage. However, there is no evidence for any transcellular transport during iontophoresis (Cullander, 1992). Similarly, electron micrographs for the iontophoretic in vivo transport of mercuric chloride in pig skin revealed that the primary pathway is via an intercellular route. Even with follicular transport, it should be noted that the final pathway is still intercellular between hair follicles and epidermal cells (Monteiro-Riviere et al., 1994). Also, the iontophoretic transport of pindolol and calcitonin through both guinea pig skin and human skin equivalent was identical. Since living skin equivalent does not contain any appendages, this suggests that an appendageal pathway is not necessary for iontophoretic transport to occur (Hager et al., 1994). Iontophoretic transport of calcein through human stratum corneum was also shown to occur through an intercellular route using scanning confocal fluorescence microscopy. Some involvement of even a transcellular path was shown, but no appendageal transport was observed (Prausnitz et al., 1996a). It has been shown that in the presence of an applied electric field, the stratum corneum lipid lamellae become more accessible to water and ions. This suggests that ion and water transport during iontophoresis is at least partly associated with stratum corneum lipid lamellae (Pechtold et al., 1996). Scanning electrochemical microscopy has been used for direct imaging of the ionic flux of Fe $(CN)_6^{4-}$ through pores of hairless mice skin. Activation of low resistance pores was shown to occur during iontophoresis, with spatial density of current-carrying pores increasing from 0 to 100–600 pores/cm^2 during the first 30–60 min of iontophoresis. The pore density reaches a quasi-steady-state value in proportion to the applied current density, with its contribution to total skin conductance increasing from 0–5 per cent to 50–95 per cent (Scott et al., 1993, 1995). Iontophoresis increases the concentration of the drug in the part of the skin which is the rate-limiting barrier during passive diffusion. Thus, for lipophilic drugs such as fentanyl, it will increase its concentration in viable skin, while for hydrophilic drugs such as thyrotrophin-releasing hormone, it will increase its concentration in the stratum corneum (Jadoul et al., 1995). This is because hydrophilic drugs have difficulty in partitioning into the stratum corneum, while lipophilic drugs have difficulty in partitioning out of the stratum corneum into viable skin.

2.4 Theoretical basis of iontophoresis

Some of the theoretical considerations developed for analysis of transdermal delivery under iontophoretic transport are discussed in this section. However, it may be noted that the techniques for electrical enhancement of percutaneous absorption, in realistic terms, constitute a rather complex area with a large number of operating variables and the results depend on the drug candidate being studied. Many of the theoretical equations for iontophoresis have been derived based on experimentation with simple ions such as sodium transport studies. These equations may not be applicable to many drugs, especially for drugs which bind to the skin or for macromolecules. The situation may be further complicated for peptide drugs which have the potential to undergo enzymatic degradation during transport through the skin and also undergo other losses such as by adsorption and self-aggregation. The basic equation also needs a modification for electro-osmotic flow which will be discussed in Section 2.6.

In general, the flux of an ionic species, i, is given as:

flux = concentration × mobility × driving force

or,

$$J_i = C_i \times m \times \text{driving force} \tag{2.1}$$

The driving force on the species i is its chemical potential gradient. Thus,

$$J_i = C_i m (-du_i/dx) \tag{2.2}$$

The thermodynamic expression for the electrochemical potential, u_i, is given as:

$$u_i = u_{i(o)} + RT \ln C_i + z_i F E \tag{2.3}$$

where $u_{i(o)}$ is the standard chemical potential and z_i is the valence of the species i, F is Faraday's constant, E is the electrostatic potential, R is the gas constant and T is the temperature in Kelvin. Assuming $u_{i(o)}$ is constant and substituting Eqn 2.3 into du_i/dx in Eqn 2.2, we get:

$$J_i = -C_i m (RT\, 1/C_i\, dC_i/dx + z_i F\, dE/dx) \tag{2.4}$$

The mobility, m, is related to diffusivity, D, as:

$$m = D_i/RT \tag{2.5}$$

Substituting Eqn 2.5 for m in Eqn 2.4, we get

$$J_i = -D_i\, dC_i/dx - z_i m F C_i\, dE/dx \tag{2.6}$$

This is a fundamental relationship, called the Nernst–Planck equation, and is widely used to describe the membrane transport of ions (Finkelstein and Mauro, 1977). Several more rigorous theoretical models have been developed for iontophoretic delivery, which have been reviewed and discussed for those interested in a more detailed treatment (Garrido et al., 1985; Keister and Kasting, 1986; Kasting and Keister, 1989; Kasting, 1992; Kontturi and Murtomaki, 1996). A better appreciation of the meaning of this equation may be achieved by considering the case of a non-electrolyte, in which the charge, $z = 0$. In this case, the Nernst–Planck equation is reduced to:

$$J = -D\, dC/dx \tag{2.7}$$

which is Fick's first law of diffusion, the same as Eqn 1.1. On the other hand, for an ion with a uniform concentration throughout the system ($dC_i/dx = 0$), the Nernst–Planck equation becomes:

$$J = -z_i m F C_i\, dE/dx \tag{2.8}$$

which is the equation for electrophoresis. The Nernst–Planck equation may be thus interpreted as implying that when a concentration gradient and an electric field both exist, the ionic flux is a linear sum of the fluxes that would arise from each effect alone. Though the permeability of ionized drugs through skin is low, it cannot be assumed to be negligible (Swarbrick et al., 1984) and thus the potential contribution of passive flux needs to be considered. An expression has been derived (Keister and Kasting, 1986; Kasting et al.,

1988) for macroscopic membranes which takes into account the lag time. This expression was derived for the effective lag time and enhancement ratio under the application of a uniform electric field to an uncharged homogeneous membrane. The enhancement ratio (E.R.), which is the ratio of the steady-state flux with applied voltage divided by steady-state flux by passive diffusion alone, was given as:

$$\text{E.R.} = J(v)/J(o) = v/1 - e^{-v}, \text{ where } v = zFE/RT \tag{2.9}$$

E.R. can also be expressed as a ratio of permeability coefficients by iontophoretic transport ($P_{\Delta E}$) over passive diffusion (P_o):

$$\text{E.R.} = P_{\Delta E}/P_o \tag{2.10}$$

Again, the assumptions made in the derivation are not likely to be valid for a complex membrane like the skin which is known to be a charged heterogeneous structure with a pH_{iso} of 3–4. This will lead to some deviation from the theory. However, the model may be a useful tool for analysing the details of iontophoresis experiments. The electric current (I) is related to the algebraic sum of the positive and negative ionic fluxes as follows:

$$I = F(\Sigma_i J_{i+} - \Sigma_k J_{k-}) \tag{2.11}$$

where J_{i+} and J_{k-} are the flux of the ionic species i^+ and k^-, respectively. By substituting Eqn 2.6 for J_i and integrating, we can derive the following expression for membrane potential:

$$\text{membrane potential} = I\,dx/F^2(u + v) + RT/F\,d(U - V)/(U + V) \tag{2.12}$$

where $U = \Sigma_i m_{i+} C_{i+}$ and $V = \Sigma_k m_{k-} C_{k-}$

m_{i+} and m_{k-} are the mobilities for ionic species i^+ and k^-, respectively. From Eqn 2.12, we can see that membrane potential is dependent on the concentration and mobility of all ions in the membrane. The first term of Eqn 2.12 is an IR drop across the membrane while the second term is a diffusion potential. Therefore, Eqn 2.12 indicates that even when the current is turned off, the diffusion potential still remains and acts as a driving force. Thus, after an iontophoresis treatment, the potential gradient in the Nernst–Planck equation (Eqn 2.6) may not be zero, but will rather depend on the concentration and mobility of all the ions in the membrane. It has been shown that the flux of an ion would be expected to remain elevated for a period of time after the current is removed since dC/dx at the membrane–receptor interface resulting from an iontophoresis treatment is significantly greater than that by passive diffusion alone.

2.5 Factors affecting iontophoretic delivery

It has been recognized for some time that there is a complex multitude of factors operating during iontophoresis (O'Malley and Oester, 1955; Zankel et al., 1959; Zankel and Durham, 1963). In order to understand the delivery profiles and to be able to use this technology commercially, it is important to understand the various formulation, electrochemical and biological factors involved in the process (Singh and Maibach, 1996). Statistical techniques such as response surface methodology can be used to minimize the number of experiments required to optimize the transdermal iontophoretic delivery of a drug under different operational conditions (Huang et al., 1995, 1996).

2.5.1 Electric current

If the transport pathways across the skin are current dependent, then an increase in current density is expected to increase the amount of drug that will be delivered (Cullander, 1992; Thysman et al., 1992; Chu et al., 1994). According to Faraday's laws of electrolysis, the transport of one molar concentration of a univalent ion requires the passage of 96 485 coulombs of electricity, if the ion has a transport number of unity. Hence, the maximum rate of transport, J_{max}, is:

$$J_{max} = (MW)\, I/96\,485 \qquad (2.13)$$

where (MW) is the molecular weight of the ion and I is the current.

The more general case for a species i is:

$$J_i = t_i\, I_T\, (MW)/z\, F \qquad (2.14)$$

where F is Faraday's constant, z is the number of charges per drug molecule (valency) and t_i is the transport number. The transport number (or transference number parameter), t_i, of an ionic species is the fraction of total applied current carried by that ionic species and may be calculated as:

$$t_i = I_i/I_T \qquad (2.15)$$

This suggests that the transference number parameter, t_i, can be assigned a fixed value for a drug ion under a set of experimental conditions even though this value could be considerably less than unity because of competition from Na^+ or Cl^- ions in the body. At low electric fields, the iontophoretic enhancement is in reasonable agreement with the predictions of the constant field model for electrodiffusion. However, at higher power levels, the drug transference number in the membrane needs to be taken into account (Kasting et al., 1988). If t_i is known, then theoretical flux can be calculated since all other parameters are known. Thus, a good prediction for each drug candidate can be made as to whether it may be a good candidate for iontophoretic delivery or not. Alternatively, the iontophoretic skin permeation (mg/h) can be predicted as follows:

$$\text{iontophoretic skin permeation} = \frac{MW \times \text{current} \times \text{current efficiency} \times 3600}{\text{molecular charge} \times \text{Faraday's constant}} \qquad (2.16)$$

The steady-state plasma levels can then be calculated by dividing the steady-state iontophoretic skin permeation by plasma clearance (Green, 1996a). As expected from Eqn 2.14, a linear dependence of the flux (J) on the total current density (I_T) applied at steady state is expected.

Several published reports support this expected result. The release of the neuropeptide angiotensin by microiontophoresis as well as its skin permeation by iontophoresis was found to be proportional to the current density applied (Harding and Felix, 1987; Clemessy et al., 1995). The drug delivery rate for various inorganic ions through various types of excised skin has been shown to be have a linear relationship with the current density (Phipps et al., 1989). Similarly, the flux of acetate ions across excised hairless mouse skin increased as the current density was increased (Miller and Smith, 1989). Flux of arginine vasopressin across excised hairless rat skin was found to be a linear function of current density and duration of application (Lelawongs et al., 1990). Thus, we can see that iontophoretic drug delivery is proportional to the applied current. Since the current is easily controlled by electronics, this technology provides a convenient means to control the rate of delivery of drugs (Sage et al., 1995). However, the current density and current intensity cannot be indefinitely increased as it will irritate and/or damage the skin, and also produce an unpleasant electrical sensation. The maximum tolerable current increases with electrode area in a non-linear fashion. While a relationship has been described in the literature (Sanderson et al., 1989; Yoshida and Roberts, 1992), the situation is understood to be much more complex. In general, 0.5 mA/cm^2 is often stated to be the maximum current density that should be used on humans without specifying the surface area.

2.5.2 *Direct current compared with pulsed current*

In direct current (DC) mode, the direction of current flow continues unchanged in one direction, while in the case of alternating current (AC), it changes periodically. Direct current may also be interrupted in a periodic manner, to produce what is called a DC pulse or periodic current. Iontophoresis is usually carried out by a continuous direct current, though the use of pulsed DC has been promoted. The waveform, current intensity, duty cycle (or on–off ratio) and frequency of periodic current may be altered. The duty cycle and the frequency of this current can be adjusted for optimum results. In addition, the current can be delivered under different waveforms, such as sinusoidal, square, triangular and trapezoidal. A pulsed waveform supposedly allows the skin to depolarize and return to its initial state before the onset of the next pulse. This is because the stratum corneum acts as a capacitor and this polarization may reduce the magnitude of a current applied as a constant current. It has also been suggested that pulsed current will be less irritating to the skin, so that patients could tolerate higher levels of current if pulsed DC at high frequency is used. Therefore, it is hoped that higher drug fluxes could be achieved with pulsed current than the equivalent DC current. Using time-dependent Nernst–Planck equations for electrodiffusion through homogeneous and structureless membranes, it has been suggested that the average flux for constant current DC compared with pulsed current is about the same for high frequencies of pulsed current (Harden and Viovy, 1996). It has been proposed that pulsed DC can result in lower skin resistance and higher drug delivery if the steady-state current during the 'on' phase of the pulse is very small and the frequency is low enough to allow depolarization of the skin during the 'off' phase. The concept of 'impact energy' based on the on-to-off ratio of pulsed DC may not apply to iontophoresis (Pikal and Shah, 1991). In an *in vitro* study on the iontophoretic transport of morphine hydrochloride across hairless mouse skin, the use of pulsed current resulted in a lower drug flux than with direct current. However, pulsed current was considered to be less damaging to the skin based on the passive flux of water across skin before and after iontophoresis for DC compared with pulsed current (Clemessy et al., 1994). Direct current has been reported to induce a higher transdermal flux than pulsed current across hairless rat skin for both fentanyl (Thysman et al., 1994b) and sufentanil (Preat and Thysman, 1993). However, polypeptides such as insulin have been reported to be better delivered by pulsed current (see Section 5.6). For delivery of thyrotropin-releasing hormone across excised rabbit pinna skin, pulsed iontophoretic flux has been reported to be higher than that obtained with continuous current. However, the frequency of the pulsed current had no significant effect on the flux (Huang and Wu, 1996a). The molecular weight may possibly be involved in the difference in efficiency between current profiles, with larger molecules being more efficiently delivered by pulsed current, though this has not been established. Differences in blood pressure reductions for DC, monophasic rectangular pulsed and monophasic trapezoidal constant current iontophoretic delivery of captopril were not found to be significant (Zakzewski et al., 1996). Similarly, Na^+ flux across excised, full-thickness, nude mouse skin by using pulsed current was the same as obtained by an equivalent continuous current, and enhanced skin depolarization at high frequency actually decreased transport (Bagniefski and Burnette, 1990). The *in vitro* iontophoretic transport of lysine and glutamic acid across hairless mouse skin has been investigated as a function of several current profiles. Pulsed DC at comparable charge delivered less than uninterrupted DC, but iontophoretic transport responded to and was well regulated by the use of different current profiles. The fluxes were directly proportional to the unipolar square-wave (positive) duty cycle. When bipolar AC (positive and negative) was used, fluxes were linear only after a threshold of 50 per cent positive duty cycle. The frequency (2.5–2500 Hz) did not affect the delivery of these small molecules (Hirvonen et al., 1995). Thus, it is not established that pulsed DC leads to better flux enhancement, contrary to what is sometimes believed and, if it is better, the mechanism is more complex than just a lowering of skin impedance (Pikal, 1995).

2.5.3 Physicochemical properties of the drug

The charge, size, structure, and lipophilicity of the drug will all influence its potential to be an iontophoresis candidate. Ideal candidates for iontophoresis should be water-soluble, potent drugs that exist in their salt form with high charge density (Gangarosa et al., 1978; Lattin et al., 1991). Conductivity experiments can be done to speculate on which drugs may be the best candidates for iontophoresis and to select the optimum pH for maximum delivery. The commonly used local anaesthetic hydrochloride salts show excellent conductivity, with the best values around pH 5 (Gangarosa et al., 1978). The conductivity of a drug can also be used to estimate the competitive transport between the drug and other extraneous ions during iontophoretic transport. In a study on the anodal transport of a model cation across excised human skin, the ratio of the specific conductance of the cation in deionized-distilled water to that of the solution applied in the donor compartment was used. A linear relationship between iontophoretic flux and specific conductance was observed, but only after a certain threshold conductivity value (Yoshida and Roberts, 1994, 1995).

Structure–transport relationships are hard to predict owing to a multitude of factors which are involved in iontophoretic delivery. The reader is referred to a comprehensive review on this subject (Yoshida and Roberts, 1992). The molecular size of the solute is a major factor determining the feasibility of iontophoretic delivery and the amount transported. The efficiency of delivery of carboxylate ions showed the following rank order: acetate > hexanoate > dodecanoate, suggesting that smaller and more hydrophilic ions are transported faster than larger ions (Miller et al., 1987; Miller and Smith, 1989). Using a series of positive, negative or uncharged solutes, the data for their iontophoretic delivery across excised human skin were best described by a linear relationship between the logarithm of the iontophoretic permeability coefficient and the molal volume, as predicted by the 'free volume' model. However, the predictions based on such models may not agree with the literature reports, in which much larger molecules have been delivered (e.g. insulin) than what would be predicted by the model (Yoshida and Roberts, 1993). In one study, the mobility of seven medications suitable for iontophoretic administration was investigated at pH 5, 7 and 9. Three positive basic drugs, acyclovir, lidocaine hydrochloride and minoxidil, were used. The other four drugs, methylprednisolone sodium succinate, dexamethasone sodium phosphate, adenine arabinoside monophosphate, and metronidazole, had a negative charge. Acyclovir was ionized by protonation of the imidazole nitrogens at positions 3 and 7 on the guanine rings. At its first pK_a of 2.27, it was 50:50 di-cation:monocation, while at its second pK_a of 9.25, it was 50:50 mono-cation:neutral species. Its mobility was observed to be best at pH 7, suggesting that mobility is best when the maximum amount of mono-cation and minimum amount of free base is available (Kamath and Gangarosa, 1995). The salt form of the drug can also play an important role in controlling delivery efficiency if a protonated drug is being used and is the primary current-carrying species in the formulation. At neutral pH, cationic drugs often exist as a mixture of protonated and unprotonated species. As the protonated drug migrates towards the skin under the electric field, an imbalance is created with more protonated drug in the boundary layer. This in turn lowers the pH in the boundary layer which results in higher proton transport. The pH may even drop below the pK of the acid. The problem will not occur with positively charged drugs which are not protonated, such as, quaternary ammonium salts. This problem can be avoided by using a weak acid salt of the drug, such as the acetate rather than the hydrochloride. For instance, the succinate salts of verapamil, gallopamil and nalbuphine all had a higher flux and current efficiency than the corresponding hydrochloride salts (Sanderson *et al.*, 1989). Non-ionized molecules will also be delivered typically by iontophoresis owing to electro-osmosis. However, if the molecule has a good passive permeability, then there may be no advantage to the use of iontophoresis. Using a series of *n*-alkanols, it was shown that iontophoresis hindered the lipoidal transport pathways. The iontophoretic enhancement values decreased linearly with increasing alkyl chain length, with the transport being even less than passive diffusion at alkyl chain lengths of greater than six (Terzo *et al.*, 1989). The upper size limit for iontophoretic delivery is not known. However, *in vitro* electrotransport of cytochrome c, a 12 400-Da protein, across human epidermis has been demonstrated (*Electrotransport: A technology whose time has come*, brochure from Alza Corporation, CA, USA, 1993). Several studies have investigated the iontophoretic delivery of insulin (see Section 5.6), which could have existed with a molecular weight as high as 36 000 since it normally occurs as a hexamer and the molecular weight of the monomer is about 6000. The molecular volume may be more important than the molecular weight as a compact folded molecule such as a globular protein may be able to pass through pores more readily than an open extended fibrous one, such as an unfolded globular or a fibrous protein. Thus, the tertiary and quaternary structure of a protein will play a role in the overall feasibility and efficiency of delivery.

2.5.4 Formulation factors

The drug concentration, pH, ionic strength and viscosity of the formulation will all affect the iontophoretic delivery of the drug. An increase in the drug concentration in the formulation will typically result in higher iontophoretic delivery, as seen with acetate ions (Miller and Smith, 1989) or metoprolol (Thysman *et al.*, 1992). Increasing the concentration will increase iontophoretic delivery up to some point, but at still higher concentrations, the flux may become independent of concentration. This could be because the boundary layer becomes saturated with the drug while the bulk donor solution is still unsaturated (Sanderson *et al.*, 1989). In order to avoid pH changes, a buffer system is usually used. The iontophoretic delivery of a drug will be reduced by these buffer ions as they will compete with the drug for carrying the current. As buffer ions are usually small, mobile and highly charged, they will usually be more efficient at carrying the current. HEPES buffer is often used in iontophoresis research as it has a high buffering capacity at pH 7.4 because it is zwitterionic at this pH and thus has reduced charge carrying ability. Addition of sodium chloride provides one primary cationic (Na^+) and one primary anionic (Cl^-) charge carrier. The pH of the buffer can be raised to 7.2–7.4 with tetrabutylammonium hydroxide. This agent is preferred for pH adjustment as it avoids the addition of small positive ions that would otherwise carry a large fraction of the charge (Burnette and Ongpipattanakul, 1987; Burnette and Bagniefski, 1988). Chloride also participates in the electrochemical reaction to form silver chloride. The buffer should be optimized for the system for which it is being used. We have used 25 mM HEPES along with about 75 mM sodium chloride and 0.02 per cent sodium azide (Banga et al., 1995b). An ethanolamine/ethanolamine HCl buffer has also been used for human studies with iontophoresis and provides the advantages of not having small mobile competitive ions and not having to add additional chloride ions (Jadoul et al., 1996). Buffers used as receptor fluids for *in vitro* studies should be deaerated prior to use by sonicating at 37°C or a slightly higher temperature to prevent bubble formation on the tissue. Alternatively, bubbles can be removed by filtration or another tested method.

An increase in ionic strength will decrease delivery as the extraneous ions will compete with the drug for the current. Since many drugs such as peptides are very potent, they are used in low concentrations so that a small amount of additives can have a rather large negative influence on delivery efficiency. Furthermore, peptides are macromolecules and have low mobility to start with. The co-ions from buffering agents are usually more mobile than the drug and will thus reduce the fraction of current carried by the drug ion. This in turn will reduce the iontophoretic delivery. Extraneous co-ions can also be introduced at the electrodes, such as the generation of hydronium and hydroxide ions when platinum electrodes are used. These ions are even more harmful to transport efficiency, their mobility being three to five times greater than small inorganic ions such as sodium, potassium and chloride (Phipps and Gyory, 1992). The ionic strength of the buffer should thus be a compromise to achieve just adequate buffer capacity to avoid pH drifts but not be too high to minimize the competition for current. An increase in solution viscosity may also decrease the iontophoretic flux by hindering the mobility of the drug (Thysman et al., 1992; Chu et al., 1994). The efficiency of drug delivery will be determined by the concentration of extraneous ions and the mobility of the drug ion in the skin relative to the mobility of these other ions. Selection of the electrode is linked to the concentration of these ions. When a lithium chloride solution was used with a platinum anode, the delivery efficiency was only 20.2 per cent owing to the generation of H^+ ions as a result of electrolysis of water. However, when a lithium nitrate solution was used with a silver anode, the delivery efficiency increased to 28.5 per cent. As discussed, ideally a chloride salt should be used. When lithium chloride was used with a silver anode, the delivery efficiency was 36.7 per cent (Phipps et al., 1989). For a discussion of optimizing the amount of halide ion in the formulation to maintain the electrochemistry with silver electrodes, the reader is referred to Section 3.4.2.

Another factor that will have a very significant influence on delivery is the pH of the formulation. The pH can determine whether or not the drug is charged or it can affect the ratio of the charged and uncharged species (Cullander and Guy, 1992). In the case of polypeptides, the type of charge is also controlled by the formulation pH relative to the isoelectric point of the polypeptide. Iontophoretic delivery of a drug may be hindered by the presence of high concentrations (> 15% v/v) of cosolvents (such as propylene glycol) in the formulation. This could be due to a decrease in the conductivity of the drug solution as well as a decrease in electro-osmotic flow (Jadoul et al., 1997).

2.5.5 Biological factors

For small ions, iontophoretic delivery may not be affected by the type of skin being used. For delivery of lithium through human, pig and rabbit skin, the iontophoretic flux was nearly identical even though the passive flux differed by more than an order of magnitude (Phipps et al., 1989). However, most drug molecules have complex structures which may interact with the skin in various ways. Thus, their iontophoretic delivery profile will have to be evaluated on a case-by-case basis. The factors involved in the selection of skin are discussed in Section 3.2. The dermal blood supply determines the systemic and underlying tissue solute absorption during iontophoresis. However, the blood supply does not appear to affect the epidermal penetration fluxes during iontophoretic delivery. This was suggested by the observation that the solute concentration in the upper layers of skin following iontophoresis was comparable in anaesthetized rats and sacrificed rats. Since the latter had no blood supply, it is presumed that the blood supply did not affect penetration through the epidermis (Cross and Roberts, 1995).

2.6 Electro-osmotic flow

If a voltage difference is applied across a charged porous membrane, bulk fluid flow or volume flow, called electro-osmosis, occurs in the same direction as the flow of counterions. This flow is not diffusion and involves a motion of the fluid without concentration gradients (Pikal, 1992). This bulk fluid flow by electro-osmosis is a significant factor in iontophoresis and was found to be of the order of μl/h per cm^2 of hairless mouse skin (Pikal and Shah, 1990a). Since skin is a permselective membrane with negative charge at physiological pH (Burnette and Ongpipattanakul, 1987), the counterions are usually cations and electro-osmotic flow occurs from anode to cathode, thus enhancing the flux of positively charged (cationic) drugs. The stratum corneum is thus often referred to as a cation permselective membrane. The major cation transported through epidermis, Na$^+$, has a transport number of 0.6, which is about twice the transport number of Cl$^-$, so that cations are transported more readily across the epidermis. The transport number of ions in skin can be calculated from potentiometric measurements, that is, measurement of the potential of a system at zero net current. The ion selectivity of the epidermis can be determined by measuring the potential difference generated when the epidermal membrane separates two chloride salts solutions of different concentrations (De Nuzzio and Berner, 1990).

Owing to electro-osmotic flow, it usually is also possible to deliver neutral drugs under the anode as the bulk fluid flow from anode to cathode can carry the neutral drug (Pikal, 1990). The term 'iontohydrokinesis' was used in early literature to describe this water transport during iontophoresis (Gangarosa et al., 1980). Electro-osmotic flow can also hinder drug flux in a situation where a negatively charged drug (anion) or a neutral drug is being delivered under the cathode. In such cases, the flux may actually increase after the current is stopped. For example, the cathodal mannitol flux through hairless mouse skin was retarded relative to passive transport owing to net volume flow in the opposite direction. The transport of mannitol increased significantly after the termination of the current (Kim et al., 1993). If the skin reverses its charge, such as at a pH below its isoelectric point, the direction of electro-osmotic flow will also reverse. The pI value of the skin surface is about 3–4, which is about the isoelectric point of keratin in the stratum corneum layer (Lin et al., 1997). As the solution pH is decreased towards 4, electro-osmotic flow decreases in magnitude and eventually will reverse direction as the charge reverses in the 'pores' of the skin somewhere between pH 3 and 4. Some positively charged peptides may actually associate with the skin to reduce or neutralize its negative charge. In these cases, the cation permselectivity of the skin may be lost, resulting in a reversal of electro-osmotic flow to the cathode-to-anode direction. This was seen for the iontophoretic delivery of nafarelin through hairless mouse and human skin (Delgadocharro and Guy, 1995; Delgadocharro et al., 1995; Bayon and Guy, 1996). The same phenomenon was also observed for another luteinizing hormone releasing hormone (LHRH) analogue, leuprolide (Hoogstraate et al., 1994). Using ^{14}C-labelled mannitol, the direction of electro-osmotic flow in the anode-to-cathode direction was first confirmed and this was dramatically reduced or reversed as nafarelin was added to the anode chamber. Iontophoresis over a 12-h period resulted in delivery of nafarelin to the receptor phase, but a significant amount associated with the skin during this time and then desorbed passively over the next 12 h (Delgadocharro and Guy, 1995). Positively charged poly-L-lysines have also been observed to reduce electro-osmotic flow in a concentration-dependent fashion. Some dependence on molecular weight was also observed. At a concentration of 10 mg/ml, the 2.7- and 8.2-kDa poly-L-lysines decreased electro-osmotic flow by five- to sixfold, while the decrease was 30-fold when the 20 kDa molecule was used at the same concentration (Hirvonen and Guy, 1997a). A series of positively charged β-blockers was also investigated for their potential to reduce electro-osmotic flow. The three most lipophilic compounds (propranolol, timolol and metoprolol) significantly reduced electro-osmotic flow while the two least hydrophobic (atenolol and nadolol) did not (Hirvonen and Guy, 1997b, c).

Electro-osmotic flow is also referred to as 'water transference number', t_w, which is defined as the number of moles of water transported per equivalent of electricity passed. Though both excised human skin and excised hairless mouse skin give significant electro-osmotic flow, water transference numbers in human skin are stated to be at least five times greater than in hairless mouse skin at the same salt concentration. However, this may not translate into higher flux for neutral species owing to the failure of water transference numbers for human skin measured at high current density (1 mA/cm^2) to extrapolate to lower current densities (Pikal, 1992). The contribution of electro-osmotic flow to drug flux enhancement during iontophoresis depends on the sign of the membrane charge, the concentration of charges in the membrane, the pore radius, the Stokes radius of the drug species, and the ionic concentration in the membrane. For high molecular weight species such as proteins, the overall flux will be low even though the relative contribution of electro-osmotic flow to flux enhancement will be large, as Stokes radius is large and diffusion is slow (Pikal, 1995). For a protein with a small net negative charge, it is possible that delivery may be higher under the wrong polarity, that is, under the anode rather than the cathode because of the higher contribution of electro-osmotic flow. For example, the flux enhancement was found to be greater for anodic delivery than for cathodic delivery for the high molecular weight anionic species, carboxy inulin and bovine serum albumin (Pikal and Shah, 1990b). For a pore radius of about 2.5 nm, the ratio of electro-osmotic flow to ionic flow is greater than unity for species larger than about 1 nm (Pikal, 1992). In another study, the flux enhancement due to electro-osmosis for model permeants with a molecular weight range of 60–504 (urea, mannitol, sucrose and raffinose) was measured across a model synthetic membrane. The flux enhancement ratio was found to depend on molecular weight, being about four times greater for raffinose than for urea. This was followed up by studying transport of urea and sucrose across human epidermal membrane. The flux enhancement ratio was observed to be three times greater for sucrose than for urea (Peck et al., 1996). The iontophoresis literature often suggests that as ions move during iontophoresis their water of hydration accounts for some or all of the electro-osmotic flow, but this concept cannot explain the reversal of direction of electro-osmotic flow when the membrane charge is reversed and it has been shown that its contribution to electro-osmotic flow is negligible (Pikal, 1992).

2.6.1 Theoretical treatment of electro-osmotic flow

Based on the above discussion, we can see that the classical Nernst–Planck equation must be modified to include an electro-osmosis component. The origin of electro-osmotic flow lies in the realm of non-equilibrium or irreversible thermodynamics. Intuitively, one may visualize electro-osmotic flow as occurring because immobile charges in the membrane require the flow of mobile counterions to maintain electroneutrality in the membrane. While theoretical models have been developed to explain electro-osmotic flow, it is unrealistic to expect that all of the complexities of a natural membrane like skin can be accurately calculated by the relatively simple models the theories must assume (Keister and Kasting, 1986; Pikal, 1992). The Nernst–Planck equation assumes that the only driving force on an ionic species 'i' is the negative gradient of chemical potential for that species alone; it neglects the potential coupling of the gradients of chemical potentials for other species 'j' with the flux of species 'i' (Finkelstein and Mauro, 1977). As discussed, an example of such coupling which is important in iontophoretic delivery, is the 'solvent drag', where the gradient of the chemical potential for water activates not only a flow of water (electro-osmosis), but also yields a flow of solute dissolved in the water. Thus, a modification to the Nernst–Planck equation (Eqn 2.6) is necessary as follows:

$$J_i = -D_i \, dC_i/dx - z_i \, m \, F \, C_i \, dE/dx \pm C_i J_v \tag{2.17}$$

where J_v is the velocity of convective flow, that is, volume flow per unit time per unit area.

2.7 Reverse iontophoresis

As the name suggests, reverse iontophoresis is the back iontophoretic extraction (by electro-osmotic flow) of a molecule from the body rather than its forward iontophoretic delivery into the body. This technique can have important applications in medical diagnostics as it can allow non-invasive sampling of biological fluids (Glikfeld et al., 1989, 1994). Reverse iontophoresis can thus be used to perform clinical chemistry without blood sampling. In addition to the advantage of being a non-invasive technique, it will allow for samples which are filtered by skin and are thus free of particulates or larger macromolecules (Merino et al., 1997). This approach has been used for iontophoretic extraction of glucose from subcutaneous tissue in an attempt to develop an alternative for the commonly used invasive and inconvenient 'finger-stick' technique. In preliminary in vitro studies using hairless mouse skin, the amount of glucose extracted at the electrodes by 2 h of iontophoresis at a current density of 0.36 mA/cm^2 was found to be proportional to the concentration of the glucose solution bathing the dermis (Rao et al., 1993). In a subsequent study in human subjects, iontophoresis at 0.25 mA/cm^2 for 60 min was applied to the ventral forearm surface to show that iontophoretic sampling of glucose is feasible. It was found that the stratum corneum contains some glucose as a product of

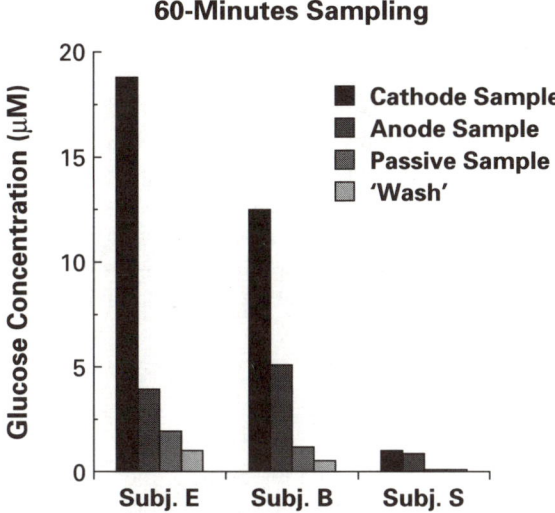

Figure 2.1 Iontophoresis sampling, at both anode and cathode chambers, of glucose (compared with passive and 'wash' controls) following current passage at 0.25 mA/cm² for 60 min. Representative data from three subjects are shown. (Reprinted from *Pharmaceutical Research*, **12** Rao *et al.*, Reverse iontophoresis: Noninvasive glucose monitoring *in vivo* in humans, pp. 1869–1873, Copyright 1995 with kind permission from Plenum Publishing Corporation.)

lipid metabolism within the skin ~~~~~~~~~~~~~~~~~~~~~ mic glucose concentration, so that ~~~~~~~~~~~~~~~~~~~~~~~~~~~~~~ be used to predict blood glucose leve~~~~~~~~~~~~~~~~~~~~~~~~~~~~~~~~~~ ithin 1 h was expected to be repres~~~~~~~~~~~~~~~~~~~~~~~~~~~~~~~~~ traction at the cathode was most ef~~~~~~~~~~~~~~~~~~~~~~~~~~~~~~~~~~ the anode (Figure 2.1). Variations a~~~~~~~~~~~~~~~~~~~~~~~~~~~~~~~~~ ic calibration may be required for ~~~~~~~~~~~~~~~~~~~~~~~~~~~~~~~~~~ s extracted by reverse iontophoresis~~~~~~~~~~~~~~~~~~~~~~~~~~~~~~~ lates or even macromolecules. T~~~~~~~~~~~~~~~~~~~~~~~~~~~~~~~ which are faced with these problen~~~~~~~~~~~~~~~~~~~~~~~~~~~~~~ A glucose monitoring device (Gluco~~~~~~~~~~~~~~~~~~~~~~~~~~~~~ cussed in Section 8.1. Detection of ~~~~~~~~~~~~~~~~~~~~~~~~~~~~~~~~ saline gel.

Several oth~~~~~~~~~~~~~~~~~~~~~~~~~~~~~~~ ble based on the concepts of revers~~~~~~~~~~~~~~~~~~~~~~~~~~~~~ as a model endogenous compound~~~~~~~~~~~~~~~~~~~~~~~~~~~~~~ o been demonstrated using chloride~~~~~~~~~~~~~~~~~~~~~~~~~~~~~~ tration of lactic acid extracted at the~~~~~~~~~~~~~~~~~~~~~~~~~~~~~~ tic acid on the dermis side (Numajiri~~~~~~~~~~~~~~~~~~~~~~~~~~~~~ flow during reverse iontophoresis h~~~~~~~~~~~~~~~~~~~~~. Since electro-osmotic flow normally takes place from anode to cathode, extraction at the cathode will be most efficient and is preferred. This will be true as long as the skin is negatively charged, which would be the case when physiological pH is used. If the pH is lowered, the charge on the skin may be neutralized as discussed before so that extraction at the anode can become feasible. The extraction was found to be enhanced if the ionic strength in the electrode chambers was reduced. The use of a pulsed current or periodic alternation of electrode polarity did not reduce the efficiency of the extraction process. The electro-osmotic flow can be increased to increase extraction efficiency at either anode or cathode by the use of some excipients such as divalent ions or EDTA in the electrode formulations, respectively. The mechanism is believed to be the modification of the net negative charge of the skin (Santi and Guy, 1996a, b). If reverse iontophoresis is to be successfully commercialized, a sensitive analytical detection method for the target material will be required as the amount extracted is likely to be very low. Though the approach can sample several materials at the same time, the need to have separate detection methods for each will pose a challenge. For the measurement of glucose, standards for day-to-day use will need to be calibrated. Also, the collection formulation will need to be optimized to improve the efficiency of extraction and the 'on-board' sensor will need to be perfected (Merino et al., 1997).

2.8 Case studies

Iontophoresis of drugs as currently used in clinical medicine for topical delivery is not discussed here. Delivery of these agents, such as corticosteroids, non-steroidal anti-inflammatory agents and lidocaine, is discussed in Chapter 4. Delivery of peptides is discussed separately in Chapter 5. In this section, the focus is on a few non-peptide drugs which have been investigated for systemic delivery for potential use in wearable patches. Several other non-peptide drugs have been investigated for systemic delivery as well and examples of isolated studies are discussed in several places in this book.

2.8.1 *Antihypertensive/cardiovascular agents*

Transdermal delivery of antihypertensive and other cardiovascular drugs offers several advantages. For nitroglycerine, the transdermal route avoids the unreliability of drug absorption from the gastrointestinal tract, precludes first-pass metabolism of the drug through the liver, and provides a continuous prolonged delivery with a 24-h patch for a drug with a half-life of just a few minutes. For clonidine, a 7-day patch has the additional advantage of minimizing the dose-dependent side effects that occur with the oral form, yet provides an adequate antihypertensive effect. Both nitroglycerine and clonidine are on the market as passive delivery patches (Shaw, 1984). Transdermal delivery of other cardiovascular agents has also been investigated (Kobayashi et al., 1993). For cardiovascular drugs, particularly for those indicated for hypertension, extended release formulations can have a significant impact on reducing patient non-compliance through less frequent dosing (Barbeau, 1995). Because of large differences in first-pass effects, the plasma concentration of propranolol can vary as much as 10- to 20-fold between individuals following oral dose. More importantly, the extended release tablet on the market cannot be conveniently used to individualize dosage for a patient or readily adjust dosage for a different indication. Iontophoretic delivery might be useful in overcoming these drawbacks and possibly in developing a self-regulated system where the delivery of propranolol is linked to blood pressure. A link to blood pressure rather than plasma concentration is sought as there is no simple correlation between dose or plasma level and therapeutic effect. This goal seems realistic as the feasibility of a non-invasive measurement of blood flow in response to propranolol or in general by the dual-beam pulsed wave Doppler technique has been established (Luca et al., 1995; Bornmyr et al., 1997). There have been several investigations on the passive transdermal absorption of propranolol and other β-blockers (Green *et al.*, 1989; Maitani *et al.*, 1993; Krishna and Pandit, 1996). These studies have attempted to use penetration enhancers such as fatty acids (Ogiso and Shintani, 1990) and alkane or alkanols (Hori *et al.*, 1992). Collagen membranes have also been investigated to fabricate transdermal delivery systems for propranolol (Thacharodi and Rao, 1996). However, the passive permeation of propranolol is unlikely to be successful as propranolol hydrochloride is a hydrophilic permeant, while skin is more permeable to lipophilic permeants. If propranolol permeates the skin as a free base, the partition coefficient for the free base is still in the order of 100-fold less than other drugs such as diazepam or indomethacin (Hori *et al.*, 1992). The use of prodrugs of propranolol has shown some promise in increasing the amounts delivered across skin (Ahmed *et al.*, 1995). Alternatively, the technique of iontophoresis can overcome these disadvantages as it expands the horizon of transdermal delivery to hydrophilic permeants. Furthermore, it may allow individualization of dosage by adjusting the current parameters. There have been only limited studies on the iontophoretic delivery of propranolol. Propranolol hydrochloride has been iontophoretically delivered *in vitro* across rat abdominal skin, and delivery was found to increase with increasing drug concentration applied, current density, duration of application and duty cycle of pulsed DC applied. Iontophoretic delivery was significantly higher than passive diffusion (Nanda and Khar, 1994). A possible barrier to the commercialization of iontophoretic delivery systems for β-blockers could be the potential for skin irritation. A high correlation has been found between the cumulative amounts of β-blockers permeating through the stratum corneum of guinea pig skin and the degree of erythema, a skin irritation reaction quantified by a chromameter (Kobayashi *et al.*, 1997).

The factors affecting the iontophoretic delivery and reversibility of skin permeability of another antihypertensive drug, verapamil, have been investigated. Following iontophoresis with pulsed DC for just 10 min, the time required for 50 per cent of the drug to desorb from the skin was as high as 8–10 h for low donor concentrations and about 20 h for high drug concentrations (Wearley *et al.*, 1989a, b). The potential of iontophoresis has also been investigated for other cardiovascular drugs such as sotalol (Labhasetwar *et al.*, 1995), and metoprolol (Okabe *et al.*, 1986; Thysman *et al.*, 1992; Ganga *et al.*, 1996). Following iontophoretic delivery of metoprolol to spontaneously hypertensive rabbits, systolic pressure was reduced from a pretreatment pressure of 126 ± 9 mmHg to a post-treatment pressure of 86 ± 11 mmHg ($P < 0.05$) within 2 h (Zakzewski and Li, 1991). The transdermal delivery of metoprolol has also been investigated by electroporation (see Section 6.5). Another hypertensive agent, captopril, is an orally effective angiotensin-I-converting enzyme inhibitor but has a relatively short elimination half-life in plasma and has been investigated for transdermal delivery (Wu *et al.*, 1996). A monophasic iontophoretic device has been investigated for delivery of captopril in rabbits. Blood pressure reduction was evident within 10 min and pronounced within 40 min. Feedback circuitry altered the pulse width of the iontophoretic signal to change the antihypertensive drug flux in order to attempt to maintain blood pressure at a target level. This attempted autoregulation failed, possibly due to the loading of the drug in a skin reservoir which provides a source of drug that continues to act long after the iontophoretic patch has been removed from the skin surface. Thus, autoregulation using simple feedback techniques may not be feasible, but predictive tools may be used to attempt autoregulation (Zakzewski *et al.*, 1996). The iontophoretic delivery of an inotropic catecholamine to dogs has been compared with intravenous infusion. Iontophoresis was found to deliver the drug to achieve the same degree of cardiac contractility and steady-state plasma concentrations as intravenous infusion (Sanderson *et al.*, 1987). The use of iontophoresis to deliver antiarrhythmic drugs from cardiac implants was mentioned in Section 1.4.3.

2.8.2 Buprenorphine

Buprenorphine is a synthetic opiate analgesic which is marketed for relief of moderate to severe pain. It also has a use in opiate addiction therapy, as it has lower abuse potential, fewer respirational effects, less physical dependence and milder withdrawal symptoms compared with methadone. Oral dosing is impractical owing to low absorption and high first-pass metabolism. Transdermal delivery of buprenorphine has been investigated and can provide therapeutic plasma levels for a sustained analgesic effect (Roy *et al.*, 1994b; Wilding *et al.*, 1996). Iontophoretic delivery could provide advantages in achieving higher plasma levels, better reproducibility and rapid onset time. However, buprenorphine is a lipophilic drug with low water solubility, which poses challenges to its iontophoretic delivery. It carries a positive charge below pH 9, with the aqueous solubility of its hydrochloride salt being 15 mg/ml at pH 4 and less than 0.1 mg/ml at pH 7. Nevertheless, in an *in vivo* study on weanling Yorkshire swine, buprenorphine was successfully delivered to achieve therapeutic doses. When delivered from the anode (0.2 mA/cm^2 for 24 h) with a concentration of 10 mg/ml, the plasma concentrations were much higher than those delivered by passive control and the dose delivered after 24 h was 0.75 ± 0.05 mg (De Nuzzio *et al.*, 1996).

2.8.3 Nicotine

Nicotine patches have been commercially available for several years and have been shown to be effective as an aid to smoking cessation (Transdermal Nicotine Study Group, 1991; Fiore *et al.*, 1994). The pharmacokinetics of the disposition of nicotine in healthy volunteers following transdermal delivery have been characterized in many studies (Kochak *et al.*, 1992; Ho and Chien, 1993; Lane *et al.*, 1993; Lin *et al.*, 1993a, b; Prather *et al.*, 1993). Nicotine has a relatively high skin permeability, so that the drug delivery patch should control the delivery, rather than allowing skin permeability to be rate limiting (Aungst, 1988). Thus, iontophoretic delivery to achieve higher blood levels is not required. However, iontophoretic delivery of nicotine may provide the advantage of achieving pulsatile drug delivery across the skin. This may be desirable as the absorption of nicotine from the lungs during smoking is in a pulsatile pattern. It has been shown (Brand and Guy, 1995) that a 1 mg 'dose' of nicotine can be delivered from a reasonably sized system within 30 min using iontophoresis. By delivering this cigarette-equivalent bolus of nicotine, the lag time of passive absorption (2–4 h) can be minimized and patients may not crave for a nicotine 'high', in particular shortly after waking.

2.8.4 Synthetic narcotics: fentanyl and sufentanil

Fentanyl and sufentanil are synthetic narcotics of the 4-anilinophenylpiperidine class. Fentanyl is a potent (80 times more potent than morphine) synthetic opioid widely used as an analgesic and as a narcotic analgesic supplement in general or regional anaesthesia due to its rapid onset, short duration of action and high potency. Owing to its extensive first-pass metabolism, it is only available for parenteral and transdermal routes. Bolus injections are not desirable as plasma concentrations will reach toxic levels and then decline rapidly owing to its short half-life. Controlled release is thus desirable and transdermal fentanyl has been widely investigated (Roy and Flynn, 1990; Donner and

Zenz, 1995; Payne et al., 1995; Simmonds, 1995) and was approved for marketing in the United States in 1991. Currently, it is the only opioid commercially available in a transdermal form and is approved for the management of chronic pain, especially cancer pain. The patch can release fentanyl for 3 days upon application, providing effective pain relief during this period. Fentanyl appears in the blood within a few hours of applying the patch, with serum levels stabilizing after 12–24 h and then staying steady over the remaining 2 days (Southam, 1995). The transdermal fentanyl patch has a slow onset of action and a long duration so that efforts are under way to develop a patch with a faster onset and shorter duration of action for the control of postoperative pain (Fiset et al., 1995; Miguel et al., 1995; Roy et al., 1996). Titration of the dose to individual patients is desirable (Korte et al., 1996) and may be accomplished by iontophoretic delivery (Thysman and Preat, 1993; Thysman et al., 1994b). Electrical and physicochemical factors affecting the iontophoretic delivery of fentanyl across hairless rat skin have been investigated. Iontophoresis was more effective from an acidic pH and the flux was found to increase with increasing current density, increasing duration of current application, and increasing drug concentration in the donor compartment (Thysman et al., 1994b). It has been shown that clinically significant doses of fentanyl citrate can be administered to humans. Analgesic doses of fentanyl were administered by iontophoresis for delivery periods of 2 h. The mean times to detectable plasma concentration were 33 and 19 min for 1- and 2-mA deliveries, with corresponding maximum concentrations being 0.76 and 1.59 ng/ml after 122 and 119 min, respectively (Ashburn et al., 1995). A wearable iontophoretic patch for delivery of fentanyl is currently under commercial development (see Section 8.1).

Sufentanil has a potency about 10 times that of fentanyl and has a similar skin permeability to fentanyl. In a study on the passive permeation of fentanyl and sufentanil through human cadaver skin, neither drug influenced the permeation of the other when they were administered concurrently (Roy and Flynn, 1990). Both fentanyl (MW 336.5) and sufentanil (MW 387.5) are relatively lipophilic molecules, with octanol/water partition coefficients of 717 and 2842, respectively (Roy and Flynn, 1988). Thus, it seems that delivery is actually hindered by the viable skin rather than the stratum corneum as clearance from the stratum corneum replaces diffusion through the stratum corneum as the rate-limiting step. In an *in vitro* study with freshly excised, abdominal, hairless rat skin, the quantity of fentanyl detected in various depths of the stratum corneum (as determined by tape stripping) following iontophoresis showed a similar distribution profile as passive diffusion. An increase in the duration of current application also did not affect the quantity of fentanyl in the stratum corneum, but larger quantities were detected in the viable tissue after iontophoresis (Jadoul et al., 1995). Iontophoretic delivery of sufentanil across hairless rat skin has been investigated. After a 2-h lag time, the iontophoretic delivery was linear with time with a flux of 219.9 ± 34.8 with direct current but only 0.8 ± 0.5 ng/cm^2 per h with passive diffusion (Preat and Thysman, 1993).

2.8.5 Miscellaneous

Azidothymidine, an anti-AIDS drug, is poorly absorbed upon oral administration and has side effects related to excessive plasma concentration immediately after parenteral administration. Thus, transdermal delivery would be useful if therapeutic levels can be delivered. This was found feasible by iontophoretic delivery based on extrapolation of *in vitro* data with hairless mouse skin (Oh et al., 1997). Transdermal delivery of capsaicin and its synthetic derivatives has been extensively investigated by passive (Fang et al.,

1996c, d; Wu *et al.*, 1997) and iontophoretic (Fang *et al.*, 1996a, b; Fang *et al.*, 1997) means by one research group. Capsaicin, the pungent principle extracted from red pepper has antinociceptive, hypotensive and hypolipidaemic activities, but its skin toxicity and burning pain sensation limits its use. Synthetic analogues of capsaicin, nonivamide and sodium nonivamide acetate, have been made available to reduce or eliminate this skin irritation (Fang *et al.*, 1996d). Iontophoresis increases the transdermal penetration flux of sodium nonivamide acetate and various factors affecting iontophoretic transport have been investigated (Fang *et al.*, 1996a, b). Another drug, methylphenidate, has also been investigated for its iontophoretic delivery. Based on the delivery of the protonated drug across excised human cadaver skin, it seems that iontophoresis can deliver therapeutic amounts of this drug (Singh *et al.*, 1997).

3

In vitro experimental techniques for iontophoresis research in the laboratory

3.1 Introduction

Several factors need to be considered in the design of iontophoresis experiments. For *in vitro* studies, these include considerations of the diffusion cells, skin source and type, electrodes, buffer and other inactive excipients in formulation, cell design, power supply, and analysis, among others. Factors affecting iontophoretic delivery including current, physicochemical properties of the drug, and formulation factors have already been discussed in Chapter 2 and are thus not discussed here. For *in vivo* studies, additional considerations involve assessment of skin damage or irritation, sensitivity of the analytical method, and patch design. Problems relating to assay sensitivity arise as the amount of drug absorbed from the skin is small and is often below assay sensitivity following dilution in the body fluids. Tracers are often used in animal studies, but the appearance of radioactivity in urine does not account for the metabolism of the drug by the skin. Thus, plasma concentrations need to be measured directly and biological response can be measured in some cases. Radioimmunoassay or ELISA kits can be used as these are very sensitive and are rapidly becoming available for many drugs. This chapter will discuss primarily the experimental techniques to conduct *in vitro* studies in the laboratory. For a good correlation between *in vitro* and *in vivo* studies, the principal barrier to transport should be the same and some considerations for iontophoretic delivery are discussed in the literature (Su *et al.*, 1994). Some novel formulations investigated will also be discussed in this chapter. Considerations of patch design for human studies are discussed in Chapter 8.

3.2 Selection of skin or membrane

3.2.1 Skin type and site

Animal skins are typically more permeable than human skin and will thus overestimate drug delivery. Hairless mouse skin was not found to be a good *in vitro* substitute for human skin when studying lidocaine penetration (Kushla and Zatz, 1991) and is also more susceptible to hydration damage than human skin (Bond and Barry, 1988). Similarly, the permeability coefficients of morphine, fentanyl and sufentanil across full-thickness hairless mouse skin were found to be one order of magnitude higher than those for human epidermis (Roy et al., 1994a). These differences can be even more in the presence of chemical enhancers. For absorption of leuprolide, the *in vitro* permeability in nude mouse skin was 10 or 100 times higher than that obtained in cadaver skin, depending on the type of enhancer that was used in the formulation. An exception is shed snake skin, which was at least 10 times less permeable than cadaver skin in this study (Lu et al., 1992). It should be noted that shed snake skin differs from human stratum corneum in that it is devoid of appendageal structures (Hinsberg et al., 1995) and it is anion selective unlike human skin which is cation selective (Hirvonen et al., 1993). Other skin types that have been used include rabbit inner pinna skin (Huang et al., 1996). Also, skin from animal species such as hairless rat often has a higher proteolytic activity than human skin. Human skin has thermoregulatory eccrine glands all over the body while other mammals including primates only have them on palmar and plantar surfaces and the perianal region. Skin of hairless rodents and even porcine skin lack eccrine glands. However, porcine skin has a similar thickness of stratum corneum as human skin, unlike hairless rodents (Cullander, 1992). Also, the hair follicle density of pig and human skin (about 11 hair follicles/cm^2) is similar, but is much higher for rat ($289/cm^2$), mouse ($658/cm^2$) or even hairless mouse ($75/cm^2$). Thus, even the use of hairless animals can be misleading as although hair shafts are lacking, rudimentary follicles are still present (Monteiro-Riviere et al., 1994). Using a series of compounds, it has been shown that the skin of miniature swine has the closest permeability characteristics to that of human skin (Bartek et al., 1972). *In vitro* studies with human skin have also been shown to correlate well with *in vivo* studies in pigs (Slough et al., 1988). In another study, the skin permeability of nicorandil was determined across excised skin samples from hairless mouse, hairless rat, guinea pig, dog, pig and human. The permeability was highest in hairless mice among the six species, while that in pig and human skin was in good agreement. It was observed that pig and human skin had similar surface lipids, barrier thickness and morphological aspects (Sato et al., 1991). However, it should be noted that the extra body fat on the pig may change the drug distribution relative to man and confound the results for *in vivo* studies (Shah et al., 1991).

The skin site is also important and care should be taken in making comparisons. For instance, the systemic bioavailability of parathion in weanling pigs was higher from the back than from the abdomen and the absorption as well as cutaneous biotransformation was altered by occlusion (Qiao and Riviere, 1995). Thus, the site and dosing method should be controlled and specified. The age of the skin may also be important. In a study on the passive permeability of water and mannitol across rat skin, age was not a significant factor for permeability but changes in dermal thickness and hair follicle depth were found to be influenced by age (Dick and Scott, 1992). These factors may affect the iontophoretic delivery of a drug across skin. The variability of skin is discussed further in Section 8.2.

3.2.2 Use of human cadaver skin

The European centre for the validation of alternative methods has recommended that there should be a concerted effort towards using human skin as the primary *in vitro* model for skin permeability studies. In case of difficulty in acquiring human skin, pig skin and hairless guinea pig skin may be acceptable alternatives (Howes et al., 1996). Since the supply of fresh and viable human skin is limited, animal skin will continue to be used.

However, human skin should be used where possible. The same should be true for electrically assisted percutaneous absorption studies. Human cadaver skin (HIV negative) can be obtained from a local skin bank, hospital or a national or international tissue source. The skin should preferably be frozen within a few hours of death, and can be supplied dermatomed to a specific thickness (typically 250–800 µm) or as full-thickness skin. Once received, the skin should be stored at a very low temperature (e.g. −80°C) and then thawed just before use. Alternatively, fresh skin can be shipped overnight if required. The source of fresh skin will be surgical procedures such as biopsies, breast skin from mastectomy or reduction, or abdominal skin from tummy-tuck surgery. Frozen human skin has been widely used in transdermal studies, for both passive and electrically enhanced delivery. The resistivity of frozen (and thawed) skin is not as high as that of human skin *in vivo*, but it is still satisfactory. The skin should be prepared carefully and electrically prescreened for defects. In a study to evaluate the electrical properties of frozen skin, the majority of tissue samples had specific resistances at 10 µA of $\geqslant$ 35 kΩ–cm^2 and sodium ion permeability coefficients comparable with human *in vivo* values. The DC current–voltage relationship for the skin was non-linear, with a decrease in resistance with increasing current values. The skin bank should be instructed to prepare the skin carefully to minimize the effect of freezing. Slow programmed freezing in the presence of a protectant such as glycerol may be required (Kasting and Bowman, 1990a). In order to ensure an intact skin barrier, other studies have used human skin samples which had at least 100 kΩ–cm^2 resistance (Prausnitz *et al.*, 1995b) or in the 20–60 kΩ–cm^2 range (Craane Van Hinsberg *et al.*, 1994). The electrical properties of fresh, excised human skin are similar, so that the use of frozen tissue in iontophoresis studies seems justified. However, fresh skin was found to be less conductive than frozen skin at low current levels, so that the driving voltage required for *in vivo* delivery devices may be somewhat higher than that expected based on *in vitro* studies with frozen skin. Also, fresh skin showed a trend towards lower sodium ion permeability, though the difference was not statistically significant. This may suggest that studies with frozen skin may not predict the electro-osmotic flow as well as those with fresh skin (Kasting and Bowman, 1990b).

3.2.3 *Cultured or living skin equivalent/skin grafts*

In recent years, a cultured or living skin equivalent of human origin has also become available and contains human dermal fibroblasts in a collagen matrix (Augustin *et al.*, 1997). This artificial skin closely resembles human skin as it has an epidermis and dermis, the former with a well differentiated stratum corneum. However, it lacks appendages such as hair follicles and sweat glands and obviously lacks the vasculature as well. A fibroblast-derived skin substitute (Advanced Tissue Sciences, Inc., La Jolla, CA, USA) was approved by the FDA in October 1997 for treatment of partial-thickness burns. Iontophoretic transport of pindolol hydrochloride, salmon calcitonin and benzyl alcohol across living skin equivalent was found to correlate well with that across guinea pig skin *in vitro* (Hager *et al.*, 1994). However, the cost and commercial availability of the skin can be drawbacks and more thorough evaluation of its potential usefulness is not available. Human skin or living skin equivalent xenografted on to immunodeficient nude mice has also been used in transdermal or basic research studies (Higounenc *et al.*, 1994; Valle *et al.*, 1996). In one study, human skin xenograft from the thigh of a 24-year-old donor was removed and stored in regular medium with antibiotics. A piece of skin measuring 2 × 1.5 cm was removed from the back of the mice, and human skin of the same size was trimmed and transplanted the same day onto anaesthetized mice by suturing into place. Mice were then used after 9 weeks when the graft was well healed (Zhang et al., 1997).

3.2.4 Skin treatment

Subcutaneous fat should be removed from skin before use. The use of full-thickness human skin can result in underestimation of *in vivo* delivery, as the blood circulation under the epidermis will normally pick up the drug so that drug transport across the full thickness of the skin is not required in the body. The *in vitro* iontophoretic transport of nafarelin across human cadaver skin varied with the type of skin used. The cumulative amount (nmol/cm^2) delivered in 24 h was 3.97 for whole skin, 28.0 for epidermis and 125 for dermis (Bayon and Guy, 1996). An applied voltage of 0.5 V across full-thickness human skin did not result in any measurable flux for two polypeptides, leuprolide and a cholecystokinin-8 analogue. However, a combination of ethanol pretreatment of skin followed by iontophoresis was found to be effective for delivery (Srinivasan et al., 1990). Epidermis can be separated prior to use, unless full-thickness skin is desired. In a typical procedure, the whole skin is immersed in water at 60°C for 45 s, at which time the epidermis can be peeled off from the dermis. Alternative methods for the separation of epidermis are also available, such as the use of EDTA or a microwave. The skin can be placed on a filter paper saturated with 0.75 per cent EDTA and kept at 37°C for 2 h, when epidermis can be separated from the rest of the skin. If desired, the epidermis can then be placed on a filter paper saturated with 0.0001 per cent trypsin solution overnight at 37°C to digest the dermis and obtain stratum corneum (Raykar et al., 1988). For separation of stratum corneum from underlying epidermis, great care is required to prevent loss of hair follicles. Samples which have holes due to loss of hair follicles will show low initial resistance (Hinsberg et al., 1995).

3.2.5 Use of synthetic membranes in iontophoresis research

The skin is a complex biological tissue and thus permeability measurements across skin tend to have a high variation. Furthermore, the application of electric current can change skin permeability, thus making delivery more unpredictable in some cases. In order to avoid these problems, various synthetic membranes have been tried in iontophoresis research. However, these membranes may not be predictive of what to expect with skin precisely because skin is a complex biological tissue and these membranes are not. The ideal membrane should be hydrophobic to mimic the lipophilic skin barrier and prevent excessive flow of water. At the same time, the membrane should also be conductive. This combination of characteristics is hard to find. Nevertheless, the use of synthetic membranes may be desirable in some situations, such as initial mechanistic preliminary studies to narrow down the number of experiments to be performed with skin or as a routine quality control for batch-to-batch variations in commercial production of iontophoretic devices (see Section 8.7). The use of two hydrophilic (Celgard-3401® and Visking 18/32®) and two hydrophobic (Celgard-2400® and Celgard-4500®; Hoechst-Celanese, Dallas, TX, USA) membranes on the passive and iontophoretic transport of salbutamol sulphate has been investigated. Slower transport was observed with the hydrophobic membranes, and it was suggested that Celgard-2400® is the best representative of human skin from these four membranes (Bayon et al., 1993). Nucleopore® (Nucleopore Corp.,

Pleasanton, CA, USA) is a synthetic membrane with essentially cylindrical, aqueous-filled pores (pore radius 75 Å; porosity 0.001). It has a polyvinylpyrrolidone-coated polycarbonate backbone with a net negative charge and a nominal thickness of 6 µm. It has been used in iontophoresis research by stacking 50 such membranes together to form a net negatively charged, random-pore network for diffusion with a resistance of about 1.5 kΩ, which is of the same order of magnitude as skin (Sims *et al.*, 1991; Hoogstraate *et al.*, 1994; Peck *et al.*, 1996; Li *et al.*, 1997). The enhanced transport of cations and anions across Nucleopore porous membranes under an applied electric field was found to be asymmetrical, possibly due to the direct effect of the field and convective solvent flow (Sims *et al.*, 1991). Thus, it seems the membrane is somewhat representative of what to expect with skin as iontophoresis is accompanied by electro-osmosis.

Drug release rates from hydrogels through cellophane membranes have also been investigated. In the absence of current, release was matrix-controlled with a linear relationship between the square root of time and amount of drug released. As current was applied, this relationship changed to a linear relationship with time for the amount of drug released (Bannon *et al.*, 1987, 1988). A microporous polyolefin membrane with hydrophilic urethane polymer-filled pores has also been used for iontophoretic delivery of the ionized drugs, dexamethasone sodium phosphate, hydrocortisone sodium phosphate and prednisolone sodium succinate, and one non-ionized drug, cortisone acetate. As expected, the electric fields interacted more efficiently with the charged molecules (Tu and Allen, 1989). Synthetic membranes may also be used as an integral part of the patch to be used on the skin. Ion exchange membranes have been investigated in this respect. In this case, the membrane should inhibit passive delivery but not delivery under an electric field. By eliminating passive release, the release of the drug would be turned on and off simply by turning the current on and off. The membrane would also protect against unintended passive absorption of the drug from abraded or compromised skin. A composite membrane containing ion exchange resins having functional groups such as sulphonic acid, carboxylic acid, iminodiacetic acid and quaternary amines in a hydrophobic polymer matrix was found to be suitable due to the creation of complex 'microporous' ion exchange pathways (Theeuwes *et al.*, 1992). Heterogeneous cation-exchange membranes have also been prepared by mixing conductive sulphonated polystyrene beads into a non-conductive silicone rubber matrix and have found use in modulating iontophoretic delivery from implantable devices (Schwendeman *et al.*, 1994). A perfluorosulphonic acid cation-exchange membrane has also been used for iontophoretic delivery of acetate ions and was stated to be a good model for skin (Miller and Smith, 1989). Iontophoretic transport of diclofenac sodium across a cellophane membrane has also been investigated (Nakhare *et al.*, 1994).

3.3 *In vitro* transdermal iontophoresis studies

Horizontal (Figure 3.1) or vertical (Figure 3.2) configuration diffusion cells are typically used for *in vitro* transdermal iontophoresis studies and will be discussed in this section. An alternative, perfused skin model is also discussed. Following a discussion of the power supply, a typical experimental set-up will be described and analytical considerations will be discussed. Alternatives to improve the methods for investigating transdermal flux will continue to evolve. For instance, a recent publication describes a new method which employs tube-shaped skin permeation cells (Moody cells) which can fit directly into standard 2-ml autosampler vials used in an HPLC system. Thus, no extra equipment is required and preliminary tests show excellent agreement with standard Bronaugh cells

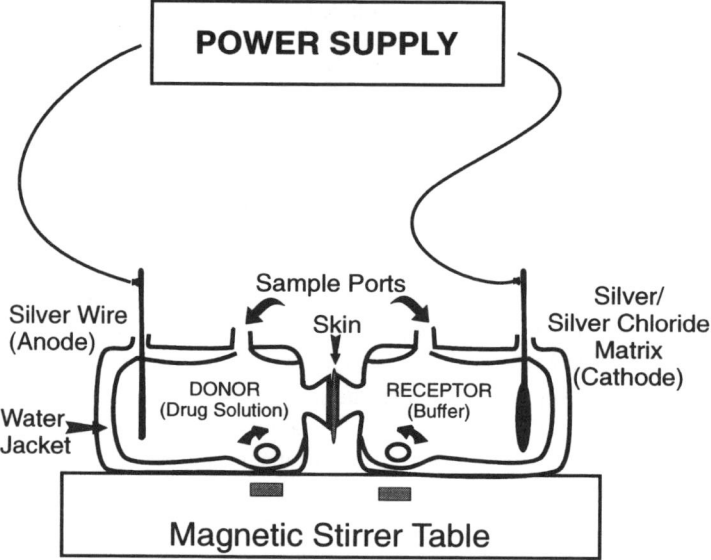

Figure 3.1 Schematic diagram of a horizontal iontophoretic diffusion cell set-up.

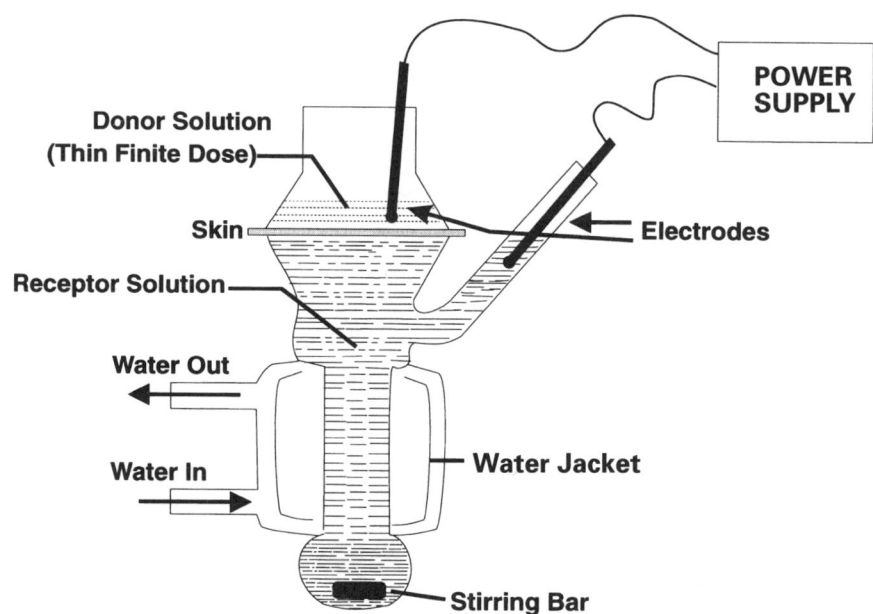

Figure 3.2 Schematic diagram of a vertical iontophoretic diffusion cell set-up.

for DEET permeation in rat skin (Moody, 1997). Other methods have also been reported (Bosman et al., 1996). The potential application of new methods to iontophoretic delivery will have to be evaluated on a case-by-case basis. The technique of microdialysis, which is based on the diffusion of analytes from the interstitial fluid through a semipermeable membrane, can be used to study the penetration of a drug in human skin layers (Ault et al., 1992). In a study with 15 healthy volunteers, microdialysis probes were inserted into defined skin layers directly under the transdermal delivery system and complete concentration versus time profiles could be obtained for nicotine. The method does have several limitations and may not apply to large (> 20 kDa) molecules or those with high lipophilicity (Muller *et al.*, 1995). Intradermal microdialysis can also be used to study the factors affecting *in vivo* iontophoretic drug delivery in a small area of the skin. The iontophoretic current does not appear to affect the performance of intradermal microdialysis. It has been found to be a useful technique to characterize iontophoretic absorption of propranolol in healthy human volunteers (Stagni *et al.*, 1997a, b).

3.3.1 Horizontal configuration

Valia–Chien or various other types of horizontal cells (see Figure 3.1) are widely used for passive, iontophoresis, electroporation or phonophoresis studies. This set-up, sometimes called the two-chambered cell, represents the so-called infinite dose technique, since the amount of drug permeating into the receptor is small relative to the total amount present in the donor. Thus, it is used for studying the mechanisms of transport and can also be used to measure absorption if the drug delivery device is intended to apply the drug to the skin at infinite dose to produce a steady-state rate of delivery. These cells should be hydrodynamically calibrated so that the studies give an intrinsic permeation rate which is independent of the hydrodynamic conditions of the cell (Tojo *et al.*, 1985). Each half-cell may have one or two ports. If two ports are present, one port serves as the sampling port while the other port can be used to insert the electrode. Excised human cadaver or animal skin or a membrane can be mounted between the two half-cells. The exposed surface area of the skin should be known for calculations, a typical value being 0.64 cm^2. An external water bath can maintain the temperature of the circulating water in the jackets at 32 or 37°C, and star-headed magnetic stirrers are used to stir the solutions in both compartments continuously.

3.3.2 Vertical configuration

Vertical diffusion cells are often used for *in vitro* permeation studies and represent the finite dose technique, as small amounts of drug can be placed on the skin. A commonly used vertical cell is the Franz cell (Figure 3.2) though other cells similar in design are also available. The donor half is exposed to room temperature (25°C), while the receptor half is maintained at 37°C, thus simulating conditions which closely approximate the *in vivo* situation. Also, the skin on the donor side is not necessarily hydrated in this set-up since ointments or other semisolid preparations can be used (Franz, 1978). Excised human cadaver or animal skin or membrane can be mounted on these vertical diffusion cells to start the experiment. Since the donor side is exposed to the environment, this cell is sometimes called the one-chamber cell.

The placement of the return electrode (in the receptor compartment) for both the designs as shown in Figures 3.1 and 3.2 may be considered as being 'inside' the skin. Any actual device, on the other hand, will place both electrodes 'on' the skin. Thus, it may be preferable to design cells which place both electrodes on the same side of the skin to simulate the *in vivo* situation. The feasibility of such a cell design was demonstrated by delivering morphine and clonidine across full-thickness hairless mouse skin. It was shown that significant lateral transport does not take place in this cell design (Glikfeld *et al.*, 1988). For evaluation of commercially available iontophoresis electrodes in clinical use, a donor chamber is not required. For such studies, a cell design has been used which uses two pieces of skin, placing one at either end of a central receptor compartment. The reservoir-type electrode filled with drug can then be placed on one side and a dispersive electrode on the other side, and samples can be taken from the central compartment (Bellantone *et al.*, 1986; Petelenz *et al.*, 1992). Commercially available horizontal cells can also be modified for a somewhat similar, single-compartment, clinically relevant design (Chen and Chien, 1994). In the case of a simple ion (benzoate), it has been observed that changes in diffusion cell configuration and placement of electrodes relative to the skin had little effect on transport (Bellantone *et al.*, 1986), perhaps suggesting that the simple diffusion cells still do have a role for *in vitro* investigations, especially for simple molecules. Flow-through systems are commercially available for either design to facilitate sampling and maintain sink conditions throughout the duration of the study. Development of automated diffusion apparatus has also been discussed in the literature (Akhter *et al.*, 1984).

3.3.3 Isolated perfused porcine skin model (IPPSF)

The IPPSF model was developed as a novel alternative animal model for dermatology and cutaneous toxicology (Riviere *et al.*, 1986) and has since been used in several transdermal delivery and/or iontophoresis investigations (Riviere *et al.*, 1991, 1996; Heit *et al.*, 1993, 1994b). These studies are discussed in other parts of this book while some details of the model itself are discussed here. Transdermal delivery of a drug involves movement from the delivery system into the skin, then into the vasculature and finally systemic disposition and pharmacodynamic effects. While most *in vitro* models can only investigate the first step, the IPPSF model can go a step further and investigate the movement into the vasculature as well. Furthermore, it provides a large dosing surface area and the convenience of continuously sampling venous perfusate. The abdominal skin of weanling pigs is typically used and a single-pedicle, axial pattern, tubed-skin flap is created following a surgical procedure. The sole vascular supply of the tube is cannulated and perfused *ex vivo* with Krebs–Ringer bicarbonate buffer containing albumin and glucose. The skin flap is then maintained in a Plexiglas chamber with controlled temperature and humidity. Viability is maintained for about 24 h (Riviere, 1996). Since the IPPSF model possesses a viable epidermis and intact vasculature, it is also useful for studies of cutaneous metabolism of drugs (Riviere *et al.*, 1996). A mathematical model to predict percutaneous absorption and subsequent disposition based on the biophysical parameters measured with the *ex vivo* perfused skin preparations has been described (Williams and Riviere, 1995). The model has been shown to predict *in vivo* serum concentrations of an iontophoretically delivered peptide (Heit *et al.*, 1993). In another study using arbutamine, a novel catecholamine, the concentration–time profiles were predicted on the basis of an IPPSF study. For two different sets of iontophoretic dosing conditions, the profiles predicted by IPPSF studies compared well with those seen after delivery to humans (Riviere *et al.*, 1992b). The use of perfused pig flap is a very attractive and useful model but one that is surgically and technically demanding, expensive, and the preparations are not usually viable for extended periods of time (Shah *et al.*, 1991; Howes *et al.*, 1996). A somewhat different study to accomplish the same objective as that of the IPPSF model cannulates the peripheral skin vein, allowing direct measurement of drug absorption *in vivo* before distribution and elimination. However, this technique results in death of the animal within a short period of time due to loss of blood. For rats, the duration of the experiment could be increased from 3 to 5 h by replacing the collected blood with blood from donor animals (Vollmer *et al.*, 1993).

3.3.4 Iontophoresis power supply

Iontophoresis power supplies that are commercially available for the topical application of drugs in physical therapy clinics are discussed separately (Chapter 4). These can also be used for *in vivo* studies in experimental animal models. The wearable patches in development also have their own miniaturized power supply, as discussed in Chapter 8. Discussed in this section are the power supplies commonly used for *in vitro* studies, which can typically power several transdermal diffusion cells at one time and allow the running of several triplicate experiments simultaneously. A Scepter® power supply built by Keltronics (Oklahoma City, OK, USA) is commercially available from PermeGear Inc. (Riegelsville, PA, USA); it operates from a PC-based software and has the capability of storing the current–voltage data on hard disk at specified intervals. Many researchers have used custom-built power supplies and some have published the circuit diagrams required to build one (Waud, 1967; Kumar *et al.*, 1992; Jaw *et al.*, 1995). A constant (DC) current of 0.5 mA/cm^2 or less should be used, as higher currents are usually not acceptable for human studies. In constant current iontophoresis, the voltage drop across the skin adjusts to keep the current constant using a two-electrode system. When there is an increase in resistance of the system, for instance due to polarization of electrodes or lack of sufficient ions to drive the electrochemistry, the power supply will respond by increasing the voltage to drive the same current through the skin. The power supply may have a limitation on the maximum voltage it can generate and thus it may be necessary to reduce the resistance in the experimental set-up. A four-electrode system is also described in the literature and is useful to test the predictions of the Nernst–Planck equation. In this four-electrode potentiostat system, the voltage drop across the skin or membrane is maintained at a constant level and the current flowing through the skin is monitored. Since the enhancement due to the applied electric field is directly proportional to the voltage drop, this system provides additional fundamental information for investigations of factors that control iontophoresis (Masada *et al.*, 1989). Using this four-electrode potentiostat with side-by-side diffusion cells, the flux enhancements for several charged model permeants across a synthetic membrane and human epidermal membranes were found to be consistent with those calculated from the modified Nernst–Planck model (Li *et al.*, 1997).

3.3.5 Typical experimental set-up

The skin should be held against the light and observed for the absence of pin holes prior to initiating the experiments. The skin should then be mounted between the donor and receptor cells. The integrity of the skin should then be checked by measuring its electrical conductivity with an ohm meter or by alternative techniques such as transport of tritiated water. If the drug has high intrinsic permeability through skin or if the iontophoretic flux is large such that a significant percentage of drug is being administered, the receptor should be replaced periodically or flow-through diffusion cells can be used. In order to maintain sink conditions for an adequate diffusion gradient and prevent back diffusion, the thermodynamic activity in the receptor should never exceed 10 per cent of that in the donor formulation. The receptor fluid must be stirred continuously and a thermostat used so as to maintain a temperature of 32°C for the skin (Howes et al., 1996). A typical in vitro experiment should last less than 7 days, with a common time period being 24–72 h. The stratum corneum obtained from human skin and mounted in a diffusion cell at 37°C has been shown to lose its barrier function at around 200 h. Immediately after separation, the passive electrical properties of the skin were similar to those in vivo but they changed within hours. At about 300 h, the magnitude of skin impedance was only about 1–2 per cent of the value immediately after separation (Pliquett and Pliquett, 1996). The duration of study will, of course, also be dictated by the therapeutic rationale for the compound being investigated. It has been suggested that protecting human epidermal membrane from physical stress is an essential element in maintaining its permeability and electrical resistance for extended periods. When supported by a porous synthetic membrane, the epidermal membrane could be used for successive passive permeability experiments over 5 days with extensive washing between permeability experiments. The advantage of successive experiments is to investigate experimental variables without factoring in skin-to-skin variability (Peck et al., 1993). For long duration experiments, suitable preservatives may have to be included in the solutions. The effect of these preservatives on the permeability of skin or the lack of their effectiveness (which may result in microbial growth) on the permeability of the skin should be carefully evaluated (Sloan et al., 1991). If the viability of the skin has to be maintained (e.g. for investigating the metabolism of the drug by the skin) during the in vitro study, fresh skin should be used with a growth medium such as Eagle's minimal essential medium or Dulbecco's modified phosphate-buffered saline supplemented with glucose. Skin viability can then be maintained for about 24 h in a flow-through diffusion cell, as measured by aerobic ^{14}C-labelled glucose utilization (Steinstrasser and Merkle, 1995). The calculations should take into account the amount of drug lost with each sampling period as the sample is typically replaced with fresh buffer. Results can be plotted as cumulative amount versus time or flux versus time. For comparing several experiments, the cumulative amount versus time curve can be used to calculate the slope during and after the iontophoresis period. This iontophoresis flux or postiontophoresis flux can then be plotted against the variable being studied, such as drug concentration or current density.

3.3.6 Analysis

For in vitro studies, samples taken from the receptor are analysed by one or more of several possible means. For studies with radiolabelled compounds, ^{14}C label in a metabolically stable position is preferable. If ^{3}H label is used, the results should be validated by an HPLC assay with a UV or radiochemical detector or other suitable stability-indicating assay. If may also be possible to lyophilize the receptor solution samples and resuspend in buffer to remove any ^{3}H label which may have exchanged with water (Prausnitz et al., 1995b). When the cumulative amount permeated is calculated, the amount lost in each sample should be taken into consideration. This can be easily set up in any commercially available spreadsheet program. The cumulative amounts can then be divided by the surface area of the skin to express results in terms of drug permeated per square centimetre of the skin, which allows more meaningful comparisons of work done in different laboratories. Cumulative amount can be plotted as a function of time, and steady-state flux (slope) can be calculated by linear regression. For partitioning studies, octanol-saturated water and water-saturated octanol should be used. Following shaking in a water bath for 24 h, the octanol–water distribution coefficient can be calculated from the ratio of drug concentration in the non-aqueous and aqueous phases. Analysis of the drug inside the skin is more tricky than analysis of the drug permeating across the skin into the receiver compartment. For *in vitro* studies, this is partly because the skin is in direct contact with the rather high drug concentration in the donor solution, which can easily confound the results. The donor compartment should be emptied and rinsed with a suitable buffer several times before the skin is removed. The skin should then be removed and rinsed thoroughly, preferably alternating rinsing and drying instead of continuous rinsing. Following rinsing, a piece of adhesive tape (such as invisible adhesive tape, 3M) can be used to strip the skin. The first stripping should be discarded. This can be followed by 15 strippings for human skin or 6 strippings for hairless rat skin to remove the stratum corneum (Howes *et al.*, 1996). However, the results should be interpreted cautiously since the quantity of stratum corneum removed can be influenced by several factors such as technique of stripping, state of skin hydration, body site and interindividual differences. It has been reported that furrows in the skin lead to further complications, with non-stripped skin in the furrows and cell layers originating from various depths due to these furrows (van der Molen *et al.*, 1997). Strips can be placed in separate scintillation vials (if radiolabelled drug was used), digested over 24 h with NaOH and then mixed with scintillation fluid for counting after neutralization with HCl (Jadoul *et al.*, 1995). Alternatively, tape strippings can be analyzed by spectrophotometric examination (Marttin *et al.*, 1996), or by placing in a combustion cone to burn in a tissue oxidizer and then analyzing the separated radionuclides by a scintillation counter (Weiner *et al.*, 1994). For non-radiolabelled compounds, proper extraction techniques will need to be developed. Autoradiography or immunohistochemical staining studies can also be performed if localization of drug within the skin is desired (Sophie and Veronique, 1991; Jadoul *et al.*, 1995). However, skin sectioning cannot normally be performed on humans unless limited to punch biopsies (Shah *et al.*, 1991). Indirect measurement of the amount in skin by measuring the amount lost from the donor is possible, but loss due to reasons other than absorption needs to be considered and it may be hard to quantify a very small difference between two very large values. Also, it should be realized that the amount in skin will vary as a function of time owing to several factors and will most likely also vary between *in vitro* and *in vivo* study as a function of time (Reifenrath *et al.*, 1991).

3.4 pH control in iontophoresis research

3.4.1 *Selection of electrode material*

The electrochemistry at the electrode–solution interface where electronic current is converted to ionic current is controlled by the electrode material used. Appropriate choice of electrodes is a factor that is critical to successful iontophoretic delivery of a drug. The electrode material in an iontophoretic device is very important as it determines the type of electrochemical reaction taking place at the electrodes. Unless special electrodes or other suitable mechanisms are used, iontophoresis is accompanied by electrolysis of water. Electrodes such as platinum, stainless steel or carbon graphite do not participate in the electrochemistry. The inert electrochemistry thus forces the water in the reservoir to become fuel for the electrochemistry (Sage, 1993). As oxidation occurs at the anode and reduction at the cathode, the following reactions take place:

$$H_2O \rightarrow 2H^+ + \tfrac{1}{2} O_2 + 2e^- \text{ (at anode)}$$

$$2H_2O + 2e^- \rightarrow H_2 + 2OH^- \text{ (at cathode)}$$

As hydrogen and hydroxyl ions are produced, this leads to a pH drop in the solution containing the anode and a rise in the pH in the solution containing the cathode. In our studies with platinum electrodes, the pH in the donor dropped to as low as 3.3, while in the receptor it rose to as high as 11.9, following 6 h of iontophoresis (Banga *et al.*, 1995b). These ions generated by hydrolysis of water are very mobile ions and may also slowly alter the conductivity of the skin. Changes in pH may lead to irritation and/or burns. Furthermore, the drug may degrade at the electrode. Significant degradation of propranolol HCl was observed when a high current was passed through platinized electrodes, and the solutions in both the receptor and donor compartments were discoloured at the end of the *in vitro* experiment (D'Emanuele and Staniforth, 1992b). The possibility of introducing metallic ions into the skin must also be carefully considered since some metals such as chromium and nickel are well known dermal allergens. Analysis of the fluid surrounding an anode made of brass or medical-grade steel by atomic absorption spectrophotometry has revealed the presence of small amounts of copper, nickel and chromium, which are known dermal allergens. The presence of these metals was found with a current density as low as 400 $\mu A/cm^2$ during 2–4 min (Linblad and Ekenvall, 1987).

3.4.2 Silver/silver chloride electrodes

A better choice of electrode material with respect to pH changes is silver for the anode and chloridized silver for the cathode. Silver/silver chloride or so-called reversible electrodes prevent pH drifts as they are consumed by the active electrochemistry, thus avoiding the use of water in the electrochemistry. However, if the current density at the silver electrode surface exceeds a limiting value for the formation of silver chloride, pH shifts might occur. At the anode, the silver oxidizes under the influence of an applied potential and when chloride ion is present it reacts to form insoluble silver chloride which precipitates on the anode surface, and an electron is released to the electrical circuit. Thus, the use of a silver anode prevents electrolysis of water, and the use of a chloride salt eliminates silver ion migration (Lattin *et al.*, 1991). Simultaneously, the silver chloride cathode is reduced, using an electron from the circuit, to silver metal that precipitates at the electrode surface and the resulting chloride ion is free to migrate into the body:

$$Ag + Cl^- \rightarrow AgCl + e^- \text{ (at anode)}$$

$$AgCl + e^- \rightarrow Ag + Cl^- \text{ (at cathode)}$$

Thus, the use of silver for the anode and chloridized silver for the cathode is desirable. The amount of chloride added to the buffer should be just enough to drive the electrochemistry, as any excess will provide avoidable competition to the drug for transport. However, enough ions must be present to ensure that the desired electrochemistry is actually occurring. The ideal case would be where the drug itself exists as a hydrohalide salt (e.g. lidocaine hydrochloride) in which case the drug itself will provide the ion for electrochemistry, resulting in high current efficiency (Sage, 1993). However, since the dose is typically low for peptides (and most transdermal delivery candidates), the fraction of current carried by the peptide is low. The counterion of the peptide thus may not be enough to drive the electrochemistry, so that some chloride (e.g. as sodium chloride) may have to be added to the formulation. This addition will adversely affect delivery efficiency since the salt has its own counterion (Na^+) which will compete for the current. Nevertheless, a hydrohalide salt will still be beneficial. Also, the silver salts of chloride or bromide are highly insoluble so that they will not migrate. However, it has been noted that most peptide drugs are available as acetate salts. Since silver acetate is much more soluble than other silver salts, there is a possibility of silver migration and skin discoloration (Sage et al., 1995).

The reaction at the anode in the absence of chloride ion may be written as:

$$Ag \rightarrow Ag^+ + e^-$$

In accordance with the above equation, we observed that in the absence of sodium chloride, the donor solution (containing the anode) became cloudy as the silver anode gradually dissolved in solution (Banga et al., 1995b). If silver ions are created, they can migrate to the skin and cause it to discolour. However, for in vitro experiments, the use of salt bridges (see Section 3.4.6) may avoid some of these problems.

The functionality of silver/silver chloride electrodes which makes them so useful in iontophoresis has been comprehensively reviewed (Cullander et al., 1993). It is related to their fast kinetics and the high conductivity of silver chloride, which permits passage of large currents with low energy input. They allow a direct relationship between electronic and ionic current, making current-controlled drug delivery possible. The electrochemical reactions take place at low potential so that undesirable secondary reactions are unlikely (the reverse is true for platinum electrodes). Silver chloride is soluble in water and its solubility product is quite low. Thus, the activity of the silver ion is very small since chloride ions are also present from other sources. However, it should be realized that electrolytically chloridized Ag/AgCl electrodes can eject colloidal AgCl particles from the surface. This may be avoided by plating the chloridized wire or using molten AgCl to form the chloridized layer. Another potential drawback is that silver ions can react with proteins containing basic or sulphur-containing amino acids, causing them to precipitate. Lastly, it should be noted that iontophoresis electrodes function under non-equilibrium conditions as an external current is passing through them (unlike reference electrodes, which are at equilibrium with their surrounding solution).

3.4.3 Preparation of silver/silver chloride electrodes

While silver/silver chloride matrix electrodes are commercially available, they can also be prepared from silver wire, if desired. It is important to use good quality, double-distilled water for the preparation of all solutions. For the formation of the silver chloride layer, two solutions are commonly used, KCl or HCl, electrolysis being carried out at moderate current densities. These solutions should be bromide free as 0.01 mole per cent of bromide can cause the silver/silver chloride electrodes to behave erratically. Silver/silver chloride electrodes can be prepared by electrolysis in a 0.1 M HCl solution at 0.4 mA/cm^2 (Miller et al., 1990). Another group lightly sanded their silver wires before placing them in 1 M HCl for 10 min at 50°C (Burnette and Ongpipattanakul, 1987). After rinsing, these were plated with AgCl by applying a current of 0.2 mA using 0.5 M KCl solution for 12 h. The electrodes were then also plated with platinum black. Using a similar technique, Ruddy and Hadzija (1995) electroplated their silver wire in 0.5 M KCl upon application of 1 mA current. Prior to electroplating, they lightly sanded the silver wire with steel wool, immersed it in 1M HCl for 30 min, rinsed it with purified water and methanol and then dried it. Following electroplating, the finished electrodes were dried and fitted with a non-conductive polyethylene sheath so as to allow only the tips to be exposed. Alternatively, silver wire has also chloridized by immersing in 133 mM NaCl solution (using Pt-cathode) for about 3 h at an applied current of 0.5 mA (Kalia and Guy, 1997). Silver/silver chloride electrodes have also been prepared by dipping silver wire into molten AgCl. After cooling, the non-coated silver wire was protected from contact with the electrolyte solution by using shrink-to-fit, salt-resistant insulating tubing (Green et al., 1991a).

3.4.4 Care of silver/silver chloride electrodes

The electrodes must be thoroughly cleaned to prevent damage to the Ag/AgCl layer by washing under flowing water immediately after use, rinsing with distilled water and air drying before storage. Continuous use of Ag/AgCl electrodes may result in a black deposit of silver chloride on top of the grey Ag/AgCl layer and can be removed by a five-to-one dilution of dilute ammonium hydroxide with distilled water. If a layer of bright silver appears, the electrode can be rechlorinated, but a homogeneous sintered Ag/AgCl layer will not exist in this case (Singh and Maibach, 1994a). Owing to the porous nature of some electrodes, it may be necessary to soak the electrodes for at least 24 h following an experiment to avoid problems of residual radioactivity or residual drug leaching into the solution in the next experiment (Pikal and Shah, 1990b). Finally, small currents should be used to minimize the depletion of AgCl present on the Ag/AgCl electrodes (Bagniefski and Burnette, 1990) and to minimize any potential damage to the skin.

3.4.5 Alternative electrode composition

An alternative to the silver/silver chloride may be the use of intercalation electrodes. These electrodes can adsorb or desorb alkali metal ions such as sodium or potassium into their structure as the electrodes are oxidized or reduced. The formation of extraneous ions and gases is minimized as a result. Suitable materials include sodium vanadate, sodium tungstate or even graphite, beta alumina, organometallic compounds, transition metal dichalcogenides, prussian blue, polyaniline, polypyrrole and polyacetyline (Untereker et al., 1996). These materials may be used as the anode or cathode and can thus deliver drugs of any charge. For example, the reaction for sodium tungstate at the anode would be:

$$Na_{1+x}WO_3 \rightarrow Na^+ + Na_xWO_3 + e^-$$

and the reaction at the cathode would be:

$$Na^+ + Na_xWO_3 + e^- \rightarrow Na_{1+x}WO_3$$

3.4.6 Alternative methods for pH control

As noted above, the addition of sodium chloride to maintain the electrochemistry for a reversible electrode will reduce the drug flux, and special device designs need to be considered. Other methods to control pH drifts are also feasible. Alternative methods may also be desirable if silver/silver chloride electrodes cause drug precipitation, as has been suggested in the case of some peptides (Lelawongs et al., 1989). Also, drugs can degrade under the electrode, as discussed. The drug may also bind to the electrode; for instance, the insulin-mimetic peroxovanadium compound has been reported to bind to the silver/

silver chloride electrodes (Brand et al., 1997). Ion exchange membranes and resins can be used for pH control. An anion exchange membrane can be used to separate the buffer from the drug solution and will prevent the buffer cations from entering the drug solution. Electrical conductivity will be maintained by the drug counterions transporting back through the ion exchange membrane (Sanderson and Deriel, 1988; Sanderson et al., 1989). Ion exchange resins have also been used for pH control under iontophoresis electrodes (Johnson and Lee, 1990; Petelenz et al., 1990). The drug ions may be attached to an ion exchange resin and, as the drug leaves the resin, the vacated site is filled by the ions of electrolysis products, so that relatively constant pH can be maintained (Petelenz et al., 1990). Alternatively, the drug need not be attached to the resin and a more generalized mechanism may be involved. The following is an example of the pH buffering action of a polymeric ion exchange resin which does not electromigrate under an applied electric field.

Two weakly acidic, cation-exchange resins, Amberlite IRP-64 and IRP-88, are mixed in a 50:50 ratio. Both resins are copolymers of methacrylic acid and divinyl benzene, with IRP-64 being the free acid form while IRP-88 is a potassium salt. For a drug with a metal ion (M^+), such as Na^+ from dexamethasone sodium phosphate, the pH buffering reaction at the cathode in presence of IRP-64 (P-COOH) is:

$$P\text{-COOH} + M^+ \rightarrow P\text{-COOM} + H^+$$

$$H^+ + OH^- \rightarrow H_2O$$

Thus the OH^- ions generated at the cathode by electrolysis are removed and the pH-control mechanism can be used to deliver negatively charged drugs. The reaction at the anode can be used for the dispersive or inactive electrode. The presence of IRP-88 (P-COOK) results in generation of K^+, which is much less mobile than the H^+ that would have been generated in absence of the resin. This should help to minimize the possibility of an electrochemical burn under the anode electrode:

$$P\text{-COOK} + H^+ \rightarrow P\text{-COOH} + K^+$$

Both silver/silver chloride as well as ion exchange resins are used in electrodes commercially available for topical delivery and are being investigated for wearable patches. Electrodes clinically being used for topical delivery are discussed separately in Chapter 4. Ion exchange resins have also been used to construct ion exchange membranes. Several electrode designs for iontophoretic delivery and for tissue stimulation are also described in the patent literature (Powers and Sisun, 1987, 1988). In some cases, the selection of the drug salt can be used for pH control. The formation of hydroxyl ions and hydrogen gas at the cathode can be avoided by selecting an anionic drug with an easily reducible counterion such as silver or copper. The counterion will be plated out or immobilized and the drug anion will be free to migrate. In another possibility, hydronium ions can be removed by protonation of the uncharged free-base form of a drug, D to form DH^+. This can be done with tertiary amines such as propranolol, nadolol and metoprolol. If an anionic drug is to be delivered, then the acid form of the drug (DH) can be used to remove the hydroxyl ions as per the following scheme (Untereker et al., 1996):

$$OH^- + HD \rightarrow H_2O + D^-$$

In order to prevent the degradation or adsorption of the drug under or on the electrodes, various researchers have used salt bridges (Figure 3.3) to apply current (Miller et al., 1990; Delgadocharro and Guy, 1995; Brand and Iversen, 1996; Brand et al., 1997).

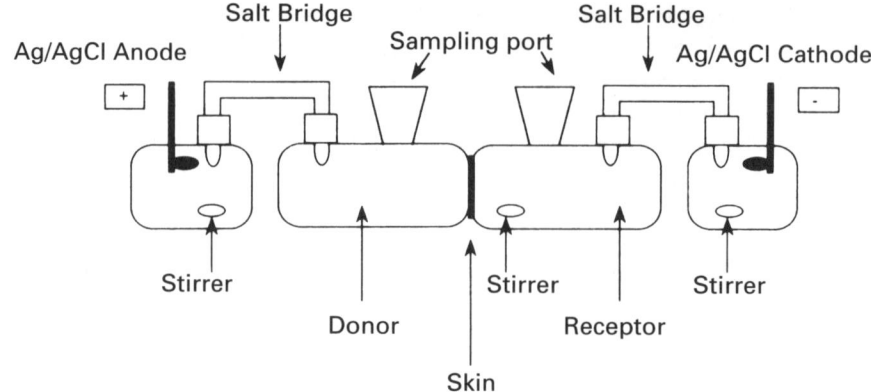

Figure 3.3 Schematic diagram of a horizontal iontophoretic diffusion cell set-up with a salt bridge. (Reprinted from *International Journal of Pharmaceutics*, **117** Delgadocharro and Guy, Iontophoretic delivery of nafarelin across the skin, pp. 165–172, Copyright 1995 with kind permission from Elsevier Science – NL, Sara Burgerhartstraat 25, 1055 KV Amsterdam, The Netherlands).

The salt bridges typically contain 3 per cent w/v agar in 1 M NaCl. These salt bridges allow electrodes to be isolated from the donor and receptor solutions.

3.5 Novel formulations for iontophoretic delivery

3.5.1 *Use of liposomes for electric enhancement of drugs*

The use of liposomes as drug carriers for topical and/or transdermal delivery has been reviewed (Rolland, 1993; Schreier and Bouwstra, 1994). Potential advantages include enhancement of drug delivery, solubilization of poorly soluble drugs, local depot for the sustained release of topically effective drugs, reduction of side effects or incompatibilities, or as rate-limiting barriers for the modulation of systemic absorption (Weiner *et al.*, 1994). It has also been suggested that topically applied liposomes can enhance the delivery of drugs into sebaceous glands (Tschan *et al.*, 1997). The potential use of liposomes to enhance the transport of conventional drugs such as oestradiol, triamcinolone acetonide, hydrocortisone, progesterone, betamethasone, methotrexate and econazole, has been widely investigated (Rolland, 1993). Liposome formulations of various lipid composition, size, charge and type all resulted in a significantly higher flux and permeability of triamcinolone acetonide through rat skin than a commercially available ointment (Yu and Liao, 1996). Dyphylline liposomes have also been delivered to the skin for potential use as a topical treatment for psoriasis (Touitou *et al.*, 1992). Econazole, an imidazole derivative useful for treatment of dermatomycosis, became the first approved dermatological product (Pevaryl® Lipogel, Cilag AG, Switzerland) available in some European countries (Naeff, 1996).

It is generally believed that intact liposomes do not traverse the skin (Schreier and Bouwstra, 1994; Hofland *et al.*, 1995), though it should be realized that such conclusions can be erroneous if based on *in vitro* studies and thus should be based on more direct studies such as microscopy. This is because the distribution coefficients of phospholipids dictate that they would stay in the skin rather than diffuse out in an aqueous environment (receptor). Liposomes may not traverse the skin but it has been suggested that they adsorb and fuse with the surfaces of the skin, and their constituents may induce changes in the ultrastructure of the intercellular lipid regions in the deeper layers of the stratum corneum, and thus produce a penetration-enhancing effect (Hofland *et al.*, 1995). The ability of liposomes to fuse with skin depends on the liposome composition and appears to be a prerequisite for skin penetration (Kirjavainen *et al.*, 1996). The size of the liposomes also appears to be important for their penetration into skin and location of the depot in the skin (Duplessis *et al.*, 1994; Natsuki *et al.*, 1996). Though liposomes are mostly prepared from phospholipids, some have been prepared from stratum corneum lipids (Wertz *et al.*, 1986; Kitagawa *et al.*, 1995) and have been used to enhance topical/transdermal delivery (Egbaria *et al.*, 1990; Yu and Liao, 1996). The use of liposomes as a carrier for peptide/protein drugs is less common but is now being exploited for various drug delivery applications (Storm *et al.*, 1995). However, it is now being recognized that liposomes present a unique opportunity to deliver the hydrophilic peptides/proteins to the skin as lipophilic vesicles. A topical formulation of liposomally encapsulated interferon was found to be effective in reducing lesion scores in the guinea pig model of the cutaneous herpes simplex virus, while application of interferon formulated as a solution or as an emulsion was ineffective (Weiner *et al.*, 1989). If lipids similar in composition to stratum corneum were used instead of phospholipids, the amount of interferon deposited in the deeper skin layers was doubled (Egbaria *et al.*, 1990). Another vesicle which appears related to liposomes but is considered to be different is called a 'transferosome'. Transferosomes are much more deformable and adaptable and can apparently bring large molecules into the skin through an intact permeability barrier. They are claimed to penetrate the skin barrier spontaneously and distribute throughout the body, possibly via the lymphatic system (Cevc *et al.*, 1995; Paul *et al.*, 1995).

As discussed, there are several reports on the use of liposomes for penetration enhancement, but the combined use of liposomes and iontophoresis has received little attention, though it could offer some additional benefits. A charge could be imparted to neutral drugs by encapsulating them in charged liposomes, followed by their iontophoretic delivery. For preparation of charged liposomes, stearylamine (SA) can be added to induce positive charge, while phosphotidylserine (PS) can be added to induce negative charge. For drugs which do carry a charge, such as peptides, liposomes could prevent or minimize their proteolytic breakdown in the skin during delivery. We were the first to report the combined use of liposomes and iontophoresis for transdermal delivery (Vutla *et al.*, 1996). In this report, we investigated the iontophoretic delivery of a liposomal formulation of a pentapeptide, Leu-enkephalin (Tyr-Gly-Gly-Phe-Leu) across human cadaver skin. Leu-enkephalin was transported across human cadaver skin using a current density of 0.5 mA/cm^2. For liposome formulations, large unilamellar vesicles of dimyristoyl phosphatidyl choline and cholesterol were prepared. The encapsulation was about 30 per cent, but was determined for each experiment, and a control solution at that concentration was then run for each iontophoresis experiment. The mean particle diameter of liposomes was 110 ± 10 nm. Following iontophoretic delivery, liposomes or their constituents were found to be present in the skin, as indicated by the presence of ^{14}C activity in the skin at the end of the experiment (Figure 3.4).

The results do not definitely establish the transport of intact liposomes, as ^{14}C cholesterol could undergo spontaneous transfer from liposomes to the skin. In these experiments, our assumption was that Leu-enkephalin is neutral as the pH equivalent to its pI was used. However, it is possible that as Leu-enkephalin enters the different skin layers, the pH changes in the microenvironment around enkephalin, resulting in a charge on the drug molecule. In this case, the charge on the enkephalin molecule may control the iontophoretic

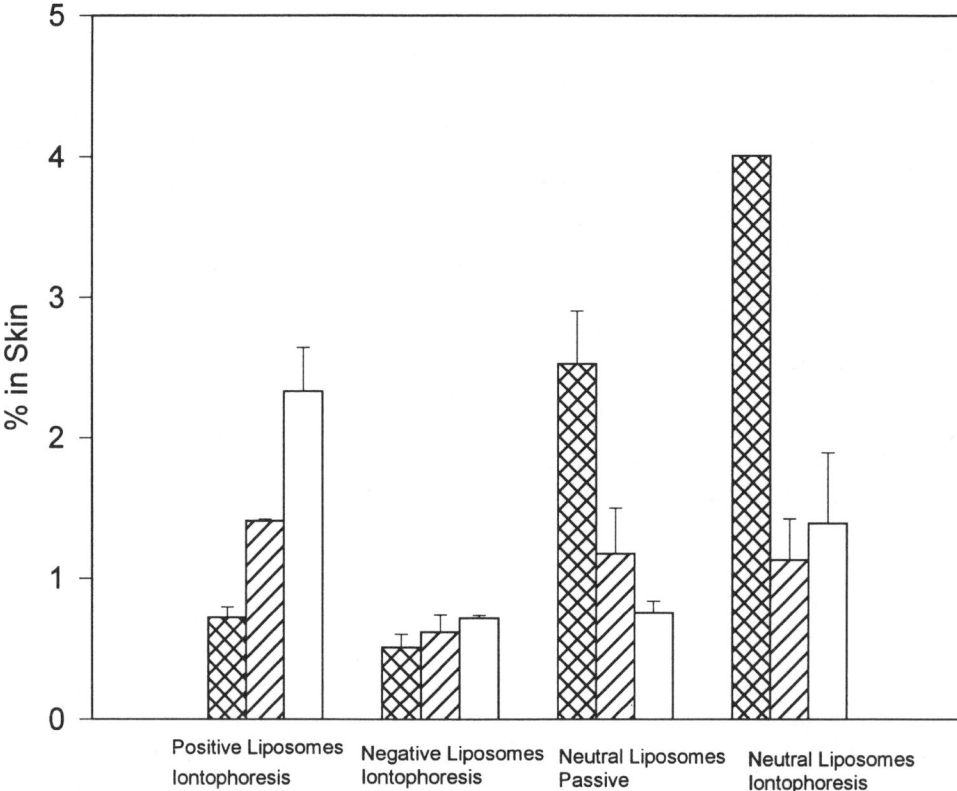

Figure 3.4 Percentage of ^{3}H-labelled [Leu^5] enkephalin (0.2 mg/ml) in the skin following iontophoresis, for control solutions (diamond patterned) and for liposomal formulations (striped). The percentage of ^{14}C-labelled liposomes in the skin (unfilled) is also plotted. The control for each group was delivered under the conditions used to deliver liposomes for that group. (Reprinted with permission from *Journal of Pharmaceutical Sciences*, **85** Vutla *et al.*, Transdermal iontophoretic delivery of enkephalin formulated in liposomes, pp. 5–8, Copyright 1996 American Chemical Society.)

flux, rather than the charge on the liposome. Also, charge interactions between liposome and enkephalin may occur at this point, rendering the complex effectively neutral. Enkephalin, when delivered iontophoretically at its isoelectric point, from liposomes carrying positive or negative charge on their surface, resulted in permeation of radioactivity which was the same or less than that of the controls when analyzed by liquid scintillation counting. The lower permeation of radioactivity with the liposome formulation may actually result from the prevention of proteolytic degradation of enkephalin by liposomes, which may result in the presence of greater quantities of the intact higher molecular weight enkephalin, compared with its degradation products, which may permeate readily. In order to assess the stability of enkephalin in these studies, an iontophoresis experiment using neutral liposomes and appropriate control was conducted and analyzed using HPLC with a radiochemical detector. Since a radiochemical detector was used for analysis, any free tritium or degradation products would not show up at the enkephalin retention time on the chromatogram. As many as five separate peaks were seen on the radiochromatography detector, upon analysis of the receptor solution following a few hours of iontophoresis. The number of degradation products could be even higher than what would appear from such a chromatogram since any cleaved fragment that did not carry the tracer would not show up on the chromatogram. When analyzed by HPLC, the skin in these experiments did not show the presence of any intact Leu-enkephalin, suggesting that the skin has metabolized all enkephalin. In studies done earlier, the radioactivity permeated continues to increase over the iontophoresis duration. However, in the HPLC study, the percentage of intact enkephalin, as determined by the radiochemical detector, starts to decline after a few hours. This could be due to the increased degradation of enkephalin as a function of time, as more and more proteolytic enzymes are leached from the skin. For the liposomal formulation, the degradation is less, indicating that liposomes are protecting enkephalin against degradation. This study had several complicating variables operating at the same time because the peptide had a charge (as a function of pH) in addition to the charge imparted on the liposomes. A simpler system would be one that uses a neutral drug in charged liposomes. We thus followed up the combined use of liposomes and iontophoresis using colchicine as a model neutral drug. The delivery of colchicine in positively charged liposomes was four to five times greater than that of plain colchicine. The composition of the lipid used to make liposomes was also found to play a role (Kulkarni et al., 1996).

3.5.2 *Hydrogel formulations for iontophoretic delivery*

Hydrogel formulations are widely used in medical devices, can provide an electroconductive base and will be more desirable for clinical use than an aqueous formulation. One advantage of a hydrogel formulation would be the ease of application as a hydrogel pad can adapt to the contours of the body. A hydrogel formulation might also be helpful in reducing skin hydration during the period of medication and might minimize the convective flow which often accompanies iontophoretic delivery. Also, when a programmable iontophoretic device is developed for therapeutic uses, a hydrogel formulation would be preferable as it can be designed as a unit dose-type drug-loaded hydrogel patch that permits the daily or weekly dosage replacement while reusing the same iontophoretic device continuously. It would also be possible to control the release rates of the drug from a hydrogel to some extent by changing the characteristics of the hydrogel formulation during synthesis. Furthermore, a hydrogel formulation may increase the skin compliance of a transdermal patch, in view of the fact that a hydrogel can absorb sweat gland secretions which, under long-term occlusion, may become irritating.

The use of p-HEMA films for the iontophoretic release of propranolol HCl has demonstrated the feasibility of using electric current to control and predict release rates of drugs from polymer membranes (D'Emanuele and Staniforth, 1991, 1992a). In a study done by the author, two types of hydrogel, polyacrylamide and p-HEMA, were synthesized (Chien and Banga, 1993) and these, along with carbopol 934 gel were used to investigate the iontophoretic release and transdermal delivery of three model peptides, insulin, calcitonin and vasopressin (Banga and Chien, 1993a). The swelling behaviour of polyacrylamide hydrogel as a function of its monomer and cross-linker concentration was studied and the results were used to synthesize hydrogel with minimal swelling. Release of a drug from a hydrogel matrix by passive diffusion has been reported to follow the Higuchi equation for matrix release. Accordingly, the cumulative amount of drug released is proportional to the square root of time, t, as per the Higuchi equation:

$$Q/A = (2 C_0 C_s D t)^{1/2}$$

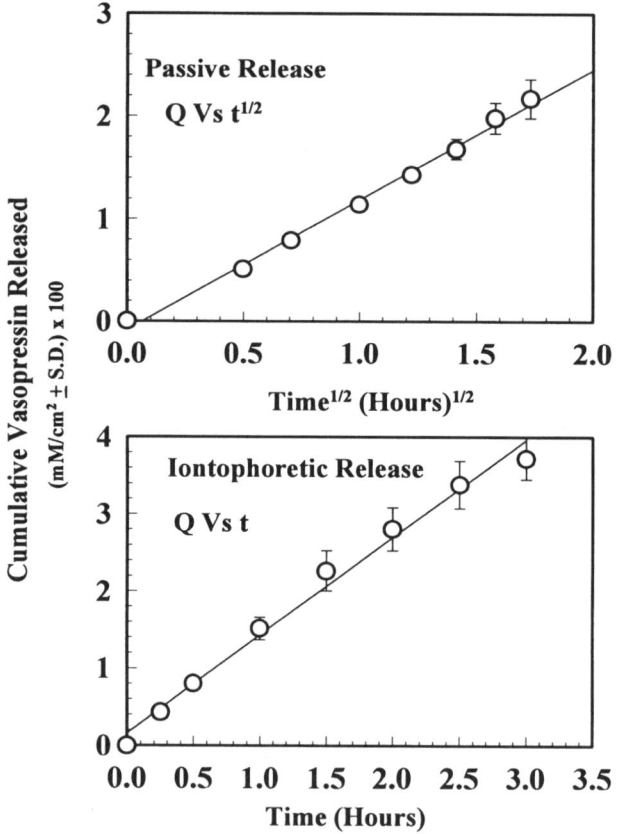

Figure 3.5 Cumulative release profile of vasopressin from polyacrylamide hydrogel, which illustrates that the release by passive diffusion follows a Q versus $t^{1/2}$ relationship, while the release by iontophoretic transport follows a Q versus t relationship. (Reprinted from *Pharmaceutical Research*, **10** (5) Banga and Chien, Hydrogel-based iontotherapeutic delivery devices for transdermal delivery of peptide/protein drugs, pp. 697–702, Copyright 1993 with permission from Plenum Publishing Corporation.)

where Q is the amount of drug released during time t, A is the surface area, D the diffusion coefficient (cm²/s), C_s the solubility of the drug in the hydrogel and C_0 is the initial drug loading in the hydrogel. For drugs which are dissolved in the polymer matrix, Baker and Lonsdale proposed the following equation:

$$Q/A = 2\, C_0\, (D\, t/\pi)^{1/2}$$

Thus, a Q versus $t^{1/2}$ relationship should be followed for the release of peptide/protein molecules from the hydrogel formulation if under passive diffusion process. This Q versus $t^{1/2}$ linearity was observed experimentally for the release of vasopressin from polyacrylamide hydrogel (Figure 3.5). However, the results in Figure 3.5 indicate that under iontophoretic transport, the release profile changed from the Q versus $t^{1/2}$ relationship to a Q versus t relationship. The same behaviour was also observed for the release of insulin, a protein molecule, from polyacrylamide hydrogel. The release kinetics of peptide/protein from hydrogel under iontophoretic transport were investigated, using insulin, under varying durations of iontophoresis application. The kinetic profiles illustrated in Figure 3.6 suggest

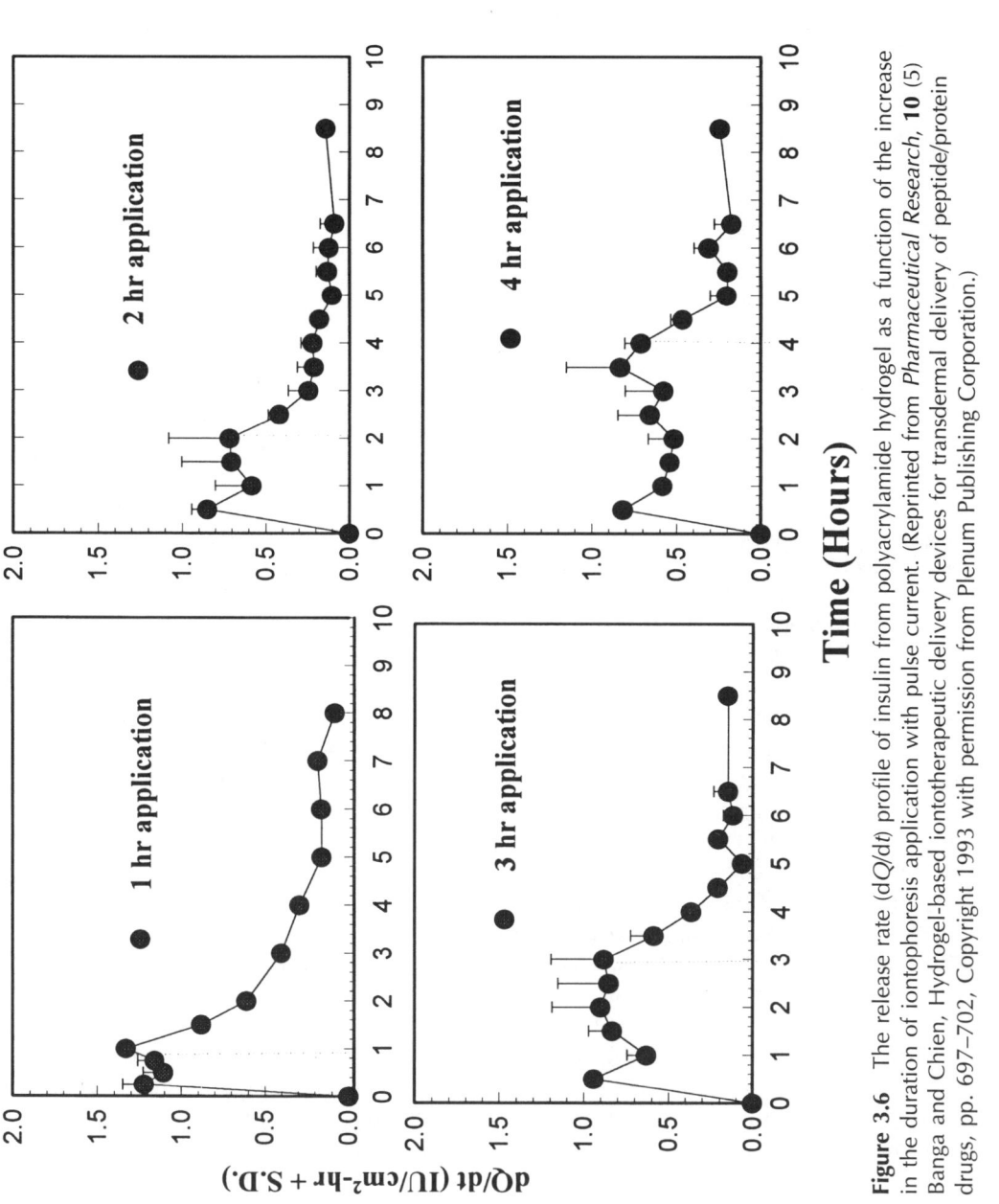

Figure 3.6 The release rate (dQ/dt) profile of insulin from polyacrylamide hydrogel as a function of the increase in the duration of iontophoresis application with pulse current. (Reprinted from *Pharmaceutical Research*, **10** (5) Banga and Chien, Hydrogel-based iontotherapeutic delivery devices for transdermal delivery of peptide/protein drugs, pp. 697–702, Copyright 1993 with permission from Plenum Publishing Corporation.)

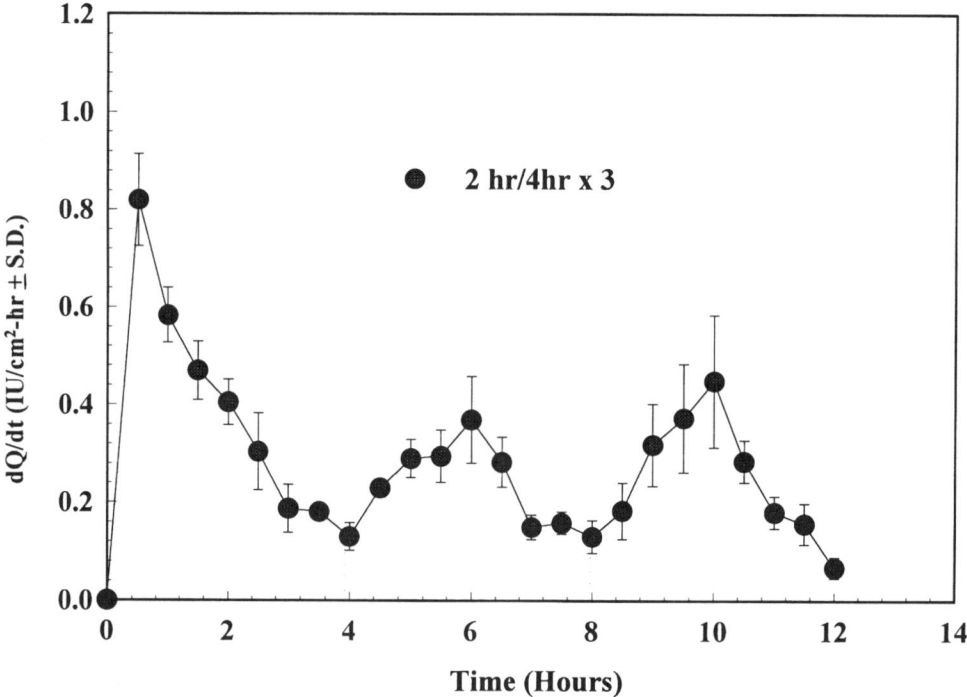

Figure 3.7 Modulation of insulin release flux from polyacrylamide hydrogel by repeated 2-h current applications at 0, 4 and 8 h. (Reprinted from *Pharmaceutical Research*, **10** (5) Banga and Chien, Hydrogel-based iontotherapeutic delivery devices for transdermal delivery of peptide/protein drugs, pp. 697–702, Copyright 1993 with permission from Plenum Publishing Corporation.)

that insulin is released at constant rate during iontophoresis application and the release rate drops as soon as the current is turned off. This behaviour was observed consistently for 1- to 4-h current application. The effect of multiple application of iontophoresis on the release flux of insulin is seen in Figure 3.7. The dQ/dt (instantaneous flux) rises as the current is applied and then declines following the termination of current application, though some burst effect is seen on first application. It then rises again on second and subsequent applications and again declines on termination of current application. Thus, the release of the peptide/protein from the hydrogel matrix can be modulated by iontophoresis application. The permeability coefficients for these peptides across the hairless rat skin were evaluated and the rank order was found to be vasopressin > calcitonin > insulin, in accordance with the order of molecular size (Banga and Chien, 1993a).

Cross-linked hyaluronic acid hydrogels have also been investigated to make an electrically responsive and pulsatile release system for negatively charged macromolecules. The solute is released when the electric field is switched off owing to the swelling of the gel and release stops when the electric field is applied again. Hyaluronic acid is a naturally occurring polysaccharide and polyelectrolyte whose cross-linked hydrogels lose water if ionic strength is high. Application of an electric field causes partial protonation of the ionized polyelectrolyte network, causing rapid deswelling of the hydrogel (Tomer *et al.*, 1995). Iontophoretic release of antimycotic agents from hydrogels through an artificial membrane using rotary disc cells has been described in the literature (Moll and Knoblauch, 1993). Gels have also been used for *in vitro* transdermal iontophoretic delivery of cromolyn sodium across hairless guinea pig skin. Gels of ionic polymers decreased the flux of cromolyn sodium but non-ionic polymers such as hydroxypropyl cellulose and polyvinyl alcohol did not affect the flux. The latter may thus be used for iontophoretic delivery from a transdermal patch (Gupta et al., 1994a). Poloxamer gels have also been used and iontophoretic release of protonated lidocaine from poloxamer 407 gels was controlled by the current density (Chen and Frank, 1997).

3.5.3 Cyclodextrins for iontophoretic delivery

The natural cyclodextrins, α-, β- and γ-, are cyclic oligosaccharides of 6, 7 and 8 glucopyranose units, respectively. The ring structure resembles a truncated core and the fundamental basis of their pharmaceutical applications is the capability of forming inclusion complexes due to the hydrophobic property of the cavity. The cavity size is the smallest ($\approx$ 5 Å) for α-cyclodextrin, followed by that of β- ($\approx$ 6 Å) and γ-cyclodextrin ($\approx$ 8 Å) (Duchene and Wouessidjewe, 1990; Szejtli, 1991b). Cyclodextrin derivatives (compared with natural cyclodextrins) exhibit higher solubility and lower toxicity by the parenteral route, while retaining their efficacy of molecular encapsulation. Hydroxypropyl-β-cyclodextrin (HPβ-CD) is a derivative with the most solubility (> 50 per cent) and the least toxicity. The necessary toxicological and human clinical data on HPβ-CD are available and its approval as an excipient is expected. Cyclodextrin–drug complexes are on the market in several countries (Szejtli, 1991a; Strattan, 1992a, b). Cyclodextrins have also been investigated as transdermal penetration enhancers (Arima et al., 1990; Vollmer et al., 1994). For example, cyclodextrins have been used to enhance the skin penetration of hydrocortisone (Loftsson et al., 1994; Preiss et al., 1995). Since cyclodextrins form inclusion complexes with drugs, it has been suggested that they may act as enhancers by interacting with components of the skin (Szejtli, 1994; Vitoria et al., 1997). Cyclodextrins can be used to bring the drug into aqueous solution to be delivered by iontophoresis (Weinshenker and O'Neill, 1991). The author's laboratory (unpublished data) has used hydroxypropyl β-cyclodextrin (HPBCD) to bring hydrocortisone into aqueous solution in relatively high concentrations and then deliver it iontophoretically (by electro-osmosis) across human cadaver skin from this cyclodextrin solution.

4
Clinical applications of iontophoresis devices for topical dermatological delivery

4.1 Introduction

Iontophoresis technology has been successfully used in clinical medicine to achieve topical delivery of drugs for several decades. Topical delivery by iontophoresis is defined as local dermatological delivery in this chapter, including delivery to the epidermis or deeper layers of the dermis. The devices and electrodes used in clinical settings for topical dermatological delivery of drugs by iontophoresis will also be discussed in this chapter. Iontophoretic wearable patches for systemic delivery are in development and are discussed elsewhere in the book, especially in Chapter 8.

4.2 Clinical applications of iontophoresis

Iontophoresis has found widespread clinical dermatological applications (Sloan and Soltani, 1986; Kassan *et al.*, 1996). Most of the clinical applications of iontophoresis currently used in physical therapy involve the use of lidocaine and dexamethasone (Costello and Jeske, 1995). However, several other drugs have been investigated for their iontophoretic delivery in clinical studies, and these will be briefly discussed. Iontophoresis also has applications in dentistry and ophthalmology, but these are beyond the scope of the present book (see Section 1.4.3). Since iontophoretic delivery takes place via an appendageal pathway, the technique may be useful for treatment of a number of follicular diseases, such as acne or androgenetic alopecia, associated with sebaceous gland activity (Illel, 1997). It should be realized that most applications discussed in this chapter are potential applications or those on which investigations have been carried out and reported in the published or patent literature. The regulatory acceptability of these applications may vary between countries.

4.2.1 *Delivery of dexamethasone*

Dexamethasone sodium phosphate at its first pK_a of 1.89 is 50:50 neutral species:mono-anion. At its second pK_a of 6.4, it is 50:50 mono-anion:di-anion. It has its maximum mobility around pH 7 (Kamath and Gangarosa, 1995). Dexamethasone, being an anionic drug, has been shown to be delivered better under the cathode, though delivery under the anode may be feasible under certain conditions (Petelenz et al., 1992) owing to electro-osmosis, which represents the water flow that accompanies iontophoresis. Electro-osmotic water movement is from the anode to cathode, thus water moves into the skin at the anode. This becomes important because dexamethasone is often delivered along with lidocaine, and the latter may be better delivered under the anode. In these cases, selection of clinical treatment protocol depends on the desired therapeutic effect by possibly switching polarity of the electrode during treatment. Switching of polarity may allow administration of drugs with opposite charge from the mixture. A major application of iontophoresis, typically utilized in physical therapy clinics, is the treatment of acute musculoskeletal inflammation. Iontophoresis or phonophoresis of dexamethasone, often in combination with lidocaine, is commonly performed in physical therapy clinics to treat local inflammatory musculoskeletal conditions such as bursitis, tendinitis, arthritis, carpal tunnel syndrome and temporomandibular joint dysfunction (Bogner and Banga, 1994). Some early studies established that iontophoresis is an effective mode for delivering dexamethasone sodium phosphate to patients for treatment of inflamed tissues (Bertolucci, 1982; Harris, 1982). Formulation and delivery considerations have been discussed more recently in the pharmaceutical literature (Allen, 1992).

In a pilot study with two monkeys, it was established that dexamethasone sodium phosphate was transported from positive electrodes placed on several sites to the underlying tissue. After sacrificing the animal, a core of tissue was obtained from under each electrode and detectable levels of dexamethasone were seen in all layers, as deep as tendinous structures and cartilaginous tissue (Glass et al., 1980). In a subsequent double-blind study on 53 patients with tendinitis at the shoulder joint, patients below 45 years of age responded to the iontophoretic administration of steroid. Patients over 45 years did not respond well to iontophoresis. However, this latter group also did not respond well to administration by local injection (Bertolucci, 1982). In a study on 18 female subjects using a Phoresor device, a single treatment of dexamethasone iontophoresis following muscle soreness was found to slow the progression of muscle soreness (Hasson et al., 1992). Iontophoretic delivery of dexamethasone has found use in rheumatoid arthritis as an alternative to oral and injected delivery of corticosteroids because of its non-invasive, non-traumatic, painless and site-specific delivery (Hasson et al., 1991; Li et al., 1996). Rheumatoid arthritis is a chronic progressive disease of the autoimmune system which is a major cause of work disability (Dawes and Fowler, 1995; Doeglas et al., 1995). Iontophoretic delivery of dexamethasone combined with a supervised exercise programme has been reported to be effective for treatment in a patient with rheumatoid arthritis (Hasson et al., 1991).

Iontophoresis of dexamethasone sodium phosphate may also offer an alternative to steroid injections for carpal tunnel syndrome, which has become an epidemic of the industrialized world. In one prospective, non-randomized study, 23 hands of 18 patients with varying occupations were treated with wrist splinting with non-steroidal anti-inflammatory medications and iontophoresis of dexamethasone using the Phoresor II system and TransQ electrodes. A success rate comparable with splinting plus injection of dexamethasone into the carpal tunnel space was achieved. In a 6-month follow-up, 83 per cent of hands failed to respond to splints plus non-steroidal anti-inflammatory medications alone. A success rate of 58 per cent was found if iontophoresis was performed in patients in failed groups who chose to proceed with this treatment (Banta, 1994). Iontophoresis or injections of steroids are also used to reduce acute inflammation of the lateral epicondyle, the bony ridge on the outer portion of the elbow, for treatment of lateral epicondylitis (tennis elbow). Tennis elbow results from performing repetitive activities such as tennis, housework, cooking, physical labour or computer operation, and iontophoresis can allow penetration of the drug to the lateral epicondyle (Liss and Liss, 1995). Iontophoresis of dexamethasone and lidocaine has also been found to be more efficient and effective in relieving pain and inflammation and to promote healing in patients with infrapatellar tendinitis than established protocols of modalities and transverse friction massage (Pellecchia *et al.*, 1994). In contrast to the studies discussed so far, some studies have suggested that iontophoretically administered dexamethasone was no more effective than saline placebo in alleviating the symptoms of patients with musculoskeletal dysfunction (Chantraine *et al.*, 1986; Reid *et al.*, 1994).

4.2.2 Delivery of other steroids

Enhanced penetration of corticosteroids such as hydrocortisone into skin may be useful for various topical or systemic diseases. For instance, topical applications can include treatment of inflammatory skin diseases (Tauber, 1994; Fuhrman *et al.*, 1997). Chemical enhancers have been used to facilitate the skin permeation of hydrocortisone (Michniak *et al.*, 1994). Iontophoretic delivery of hydrocortisone has also been investigated. The iontophoretic flux, even in the presence of surfactants, was low owing to the relatively poor aqueous solubility of hydrocortisone and only limited capacity of surfactants to solubilize poorly soluble drugs (Wang *et al.*, 1993). Salt forms such as hydrocortisone sodium succinate and hydrocortisone sodium phosphate will result in a higher iontophoretic flux, but these salt forms tend to be unstable. The salt forms of hydrocortisone have ester linkages at C21 which may undergo hydrolysis during iontophoresis, especially when the pH of donor and receptor solutions shifts, such as when platinum electrodes are used (Seth *et al.*, 1994). Successful iontophoretic delivery of other steroids such as prednisolone through skin has also been reported, though a reservoir was observed to be formed in the skin. Iontophoresis of prednisolone through nail was also found to be feasible after soaking the nail in soapy water for 5 min to make it conductive (James *et al.*, 1986). Iontophoretic delivery of steroids has also been investigated for treatment of Peyronie's disease (Rothfeld and Murray, 1967). However, dexamethasone remains the drug of choice because of its much greater anti-inflammatory effect.

4.2.3 Delivery of lidocaine

The successful use of iontophoresis for local anaesthesia, using lidocaine as the drug of choice, has been reported in the dental, ENT, ophthalmic and physical medicine literature (Henley, 1991). Iontophoresis of lidocaine in the external ear of patients produces anaesthesia of the tympanic membrane and can be useful in permitting painless execution of myringotomy or minor surgical procedures (Comeau *et al.*, 1973; Echols *et al.*, 1975; Ramsden *et al.*, 1977; Comeau and Brummett, 1978). Lidocaine exists as its base (MW 234.3) or hydrochloride salt (MW 270.8), is very stable in solution and has an acid dissociation constant of 7.84 (determined potentiometrically) at 25°C (Groningsson *et al.*, 1985), though a recent study determined the pK_a as 7.16 by precision conductometry (Sjoberg *et al.*, 1996). At physiological pH, lidocaine hydrochloride is a positive drug owing to protonation of the amine nitrogen (Kamath and Gangarosa, 1995). The ionization conditions for iontophoretic delivery of lidocaine have been recently described (Karami *et al.*, 1997). A basic study on the factors affecting iontophoretic delivery of lidocaine across human stratum corneum has been reported (Siddiqui *et al.*, 1985). Iontophoretic delivery of lidocaine is often performed for local anaesthesia or in combination with dexamethasone for the treatment of acute musculoskeletal inflammation. Lidocaine will be about 95 per cent ionized at a pH of about 6 and thus can be delivered under the positive electrode. Iontophoresis of lidocaine has been reported to result in high concentrations of lidocaine in underlying tissues when compared with passive application to rat skin (Singh and Roberts, 1993).

Lidocaine anaesthesia induced by iontophoresis under positive electrode using a Phoresor device was compared with injection or simple topical application in 27 subjects. It was found that the anaesthesia induced by iontophoresis of lidocaine has a shorter duration and lower clinical efficacy than tissue infiltration via injection, but significantly better outcomes than topical application of lidocaine. The iontophoresis-induced local anaesthesia was effective for approximately 5 min (Russo *et al.*, 1980). Iontophoresis of lidocaine has also been compared with a commercially available (EMLA, Astra) topical anaesthetic cream. EMLA cream (lidocaine 2.5 per cent and prilocaine 2.5 per cent) is an emulsion in which the oil phase is a eutectic mixture of lidocaine and prilocaine in a ratio of 1:1 by weight. A greater degree of anaesthesia was reported by iontophoresis of lidocaine using Lectro Patch® for 30 min than EMLA left on the skin for 30 or 60 min (Greenbaum and Bernstein, 1994). The absorption of lidocaine by the cutaneous microvasculature may be controlled by the co-iontophoresis of vasoactive compounds. For example, co-iontophoresis of the vasodilator tolazoline has been found to increase *in vivo* systemic absorption of lidocaine, while the vasoconstrictor noradrenalin decreased the iontophoretic absorption of lidocaine (Riviere *et al.*, 1991). Conversely, a lower systemic absorption can lead to a higher cutaneous depot. Thus, co-iontophoresis of noradrenalin resulted in increased concentration of lidocaine in skin up to a depth of 3 mm, while tolazoline decreased tissue concentration of lidocaine (Riviere *et al.*, 1992a). A pharmacokinetic model describing percutaneous absorption of iontophoretically driven, topically applied lidocaine in isolated porcine flap has been described and is capable of predicting 8-h lidocaine absorptions and documents vascular effects of co-iontophoresed vasoactive compounds (Williams and Riviere, 1993).

Iontophoresis has also been used in experimental clinical studies to investigate the effect of the vasodilators acetylcholine and sodium nitroprusside on cutaneous perfusion (Morris and Shore, 1996). Iontophoresis of lidocaine may be useful in a variety of situations, including providing local anaesthesia during pulsed dye ablation of port-wine stains (Kennard and Whitaker, 1992), painless cauterization of spider veins (Bezzant *et al.*, 1988), painless injections (Petelenz *et al.*, 1984) and painless venipuncture (Arvidsson *et al.*, 1984; Zeltzer *et al.*, 1991). The potential use of iontophoretic delivery of lidocaine for acute wound healing has also been investigated (Ernst *et al.*, 1995). Lacerations were created in the back of guinea pigs and then lidocaine was delivered using a Lectro Patch® and compared with a control group receiving lidocaine via injection. Significantly more granuloma and granulation tissue was formed in the iontophoresis group on day 10 following treatment. The results suggested that iontophoresis may have additional potential advantages for laceration repair. A combination of 2 per cent lidocaine HCl and 1:100 000 adrenalin is also approved (Iontocaine®) for dermal iontophoresis with the Phoresor system (Iomed, Inc., UT, USA). This is the first drug–iontophoresis device combination to be approved by the FDA (Guy, 1996; Merino *et al.*, 1997). This combination can achieve dermal anaesthesia to depths of up to 1 cm in about 10 min (NDA No. 20–530 filed by Iomed, Inc.). The formulation is produced for Iomed by Abbott Laboratories (IL, USA). Adrenalin is a vasoconstrictor and as it is positively charged, it may be administered along with lidocaine under a positive electrode to prolong the anaesthetic activity of lidocaine (Costello and Jeske, 1995).

4.2.4 Non-steroidal anti-inflammatory drugs (NSAIDs)

NSAIDs may cause gastric mucosal damage which may result in ulceration and/or bleeding. For some NSAIDs such as ketoprofen, these adverse gastrointestinal effects occur in about 10 to 30 per cent of patients receiving ketoprofen, and may be severe enough to require discontinuance of drug therapy in about 5 to 15 per cent of patients. These adverse side effects could be avoided by topical administration for local or systemic effects. This has prompted investigations on the percutaneous absorption of non-steroidal anti-inflammatory agents such as ketorolac acid (Roy and Manoukian, 1994; Roy et al., 1995a, b), indomethacin (Liu et al., 1995), diclofenac (Famaey et al., 1982; Vecchini and Grossi, 1984; Obata et al., 1993), ibuprofen and naproxen (Irwin et al., 1990), piroxicam (Santoyo et al., 1995), and flurbiprofen (Uchida et al., 1995), and initial results have been promising. Using a series of NSAIDs, the most suitable candidate for transdermal delivery has been stated to be ketorolac. Based on an in vitro passive permeation study with human skin, ketorolac will provide plasma concentrations which at steady state would be nearest to the therapeutic concentration (Cordero et al., 1997). Though the bioavailability from a topical site is expected to be much lower than oral absorption, topical application could result in a high local drug concentration at the diseased site. Furthermore, since the blood supply to diseased sites may be reduced, the dosage requirements for topical delivery could be less than those for oral delivery. While some potent NSAIDs may have sufficient permeability to be useful in topical formulations (Tessari et al., 1995), others may require enhancement of percutaneous absorption. The relatively low percutaneous penetration of anti-inflammatory agents may be increased by iontophoresis or phonophoresis. In a study on the delivery of indomethacin to pigs, iontophoresis (0.1 mA/cm^2) increased the maximum indomethacin blood levels from 32 (controls) to 82 ng/ml, and urinary excretion over 5 h from 29.4 to 181.1 ng/cm^2 of treated skin area, while sonophoresis did not improve indomethacin absorption. This study also looked at delivery of indomethacin to seven human subjects. Using skin of the back (1380 cm^2), the plasma levels increased from 43 ng/ml (controls) to 221 ng/ml by 1 h of iontophoresis. The urinary excretion over 5 h increased from 18.1 (controls) to 97.6 ng/cm^2 (Pratzel et al., 1986).

Clinical studies for iontophoretic delivery of NSAIDs are based on relief of clinical symptoms (Famaey et al., 1982; Vecchini and Grossi, 1984; Garagiola et al., 1988; Saggini et al., 1996) and there are very few studies which actually measure the level of drug in tissue or blood following iontophoresis. In a collaborative study done by the author on iontophoretic delivery of ketoprofen to human subjects (Panus et al., 1996, 1997), iontophoresis was conducted using the Dupel® (Empi, St Paul, MN) iontophoretic system and medium electrodes (14.2 cm^2), at a current density of 0.28 mA/cm^2 (4 mA/ 14 cm^2). Ketoprofen (300 mg/ml) in phosphate buffer (pH 7.4) with 20 per cent ethanol was sterilized before use and delivered at the volar surface of the wrist under the cathode. The treatment parameters were set as a maximum current of 4 milliamperes (mA) for 40 min resulting in a total of 160 mA*min. Forty minutes of cathodic iontophoresis were

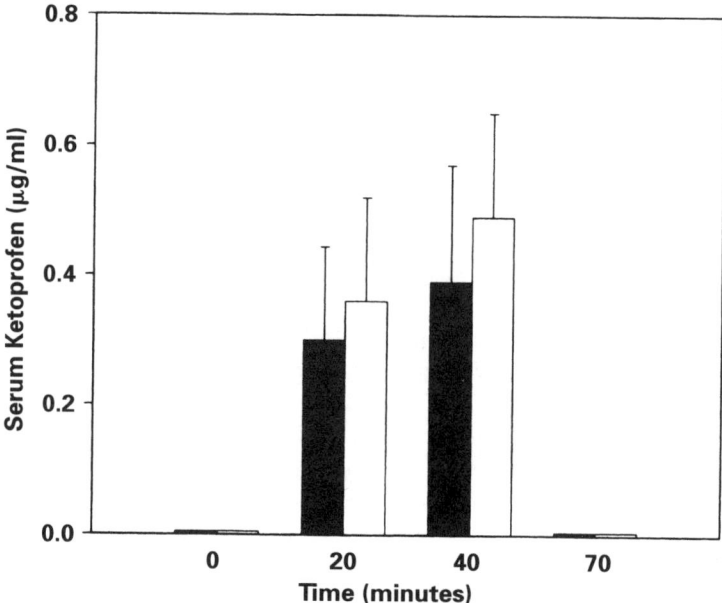

Figure 4.1 Detection of ketoprofen *in vivo* in healthy volunteers during and following 40 min of cathodic iontophoresis at 0.28 mA/cm^2. Transcutaneous delivery of individual R (■) and S (□) enantiomers was determined in the serum ($n = 3$). (Reprinted from *Journal of Controlled Release*, **44** Panus et al., Transdermal iontophoretic delivery of ketoprofen through human cadaver skin and in humans, pp. 113–121, Copyright 1997 with kind permission of Elsevier Science – NL, Sara Burgerhartstraat 25, 1055 KV Amsterdam, The Netherlands.)

required for measurable ketoprofen levels in all subjects. The ketoprofen was detected in the arm ipsilateral to the iontophoresis (Figure 4.1). The total ketoprofen concentration in the arm contralateral to the iontophoresis was below detectability 30 min following the termination of the iontophoresis. Accumulation of conjugated and unconjugated ketoprofen in the urine was also examined, with 75 per cent of the drug being excreted 8 h after iontophoresis. In this investigation, no adverse effects of the current were observed in any of the subjects following iontophoresis. A mild tingling sensation was felt by the subjects as the current was increased. A slight redness of the skin was observed under the electrodes, and this was resolved within a few hours (Panus et al., 1997). When the same dose (160 mA∗min) was applied as a 2 mA current for 80 min (instead of 4 mA for 40 min), the resulting serum levels were reduced 10-fold while the total ketoprofen and conjugates excreted in the urine were reduced to one-half. The reasons for this difference are not clear but could be related to changes in skin resistance and its recovery, local hyperaemia, or formation of a skin depot as a function of magnitude and duration of current application. This study suggests that the use of total exposure (mA∗min) to represent drug dosage may be inaccurate, contrary to what is commonly used in physical therapy clinics (Panus et al., 1996). The use of mA∗min to represent dose may also result in problems due to variation in delivery from electrodes of different manufacturers, differences in formulation, and so on. In these ketoprofen studies done by us, there was no significant difference in the *R* and *S* ketoprofen enantiomer concentrations in either the serum or the urine. Stereoselectivity in skin permeation has been suggested but it seems that no evidence is yet available. Furthermore, any stereospecific interactions may be subtle and thus may be overwhelmed by other factors such as the variability in permeation through the complex structure of the skin (Heard and Brain, 1995).

4.2.5 Treatment of hyperhidrosis

Hyperhidrosis is a condition of excessive or abnormal sweating, far higher than what is required to maintain constant body temperature and not related to exercise or resulting from another underlying cause. It can occur in feet (plantar hyperhidrosis), armpits (axillary hyperhidrosis) or hands (palmar hyperhidrosis). Generally, palmar hyperhidrosis is most distressing as hands are much more exposed in social and professional activities than other parts of the body. For hyperhidrosis without known cause (primary or essential hyperhidrosis), iontophoresis is often tried if antiperspirants do not lead to desired results (Dobson, 1987). Use of iontophoresis for hyperhidrosis was documented as early as 1952, but was studied more systematically in human subjects in the 1980s when it was reported to be effective for palmar hyperhidrosis (Levit, 1980; Stolman, 1987). The treatment consists of applying a direct current to the skin of sufficient magnitude and duration apparently to obstruct sweat glands, though the exact mechanism is not known. One study has actually examined the sweat glands from the palm of a patient before and after tap water iontophoresis by light and electron microscopy. No changes in structure were seen and it was suggested that the obstruction of sweat glands is not responsible for the effectiveness of iontophoresis to treat hyperhidrosis (Hill *et al.*, 1981). Current is typically applied in 10–20-min sessions, which need to be repeated two or three times per week, followed by a maintenance programme of treatments at 1–4-week intervals, depending on the patient's response.

The General Medical Company also offers a device (Drionic®) for treatment of hyperhidrosis and the device is available in different configurations for underarm use or use with hand or foot. The Drionic device has been reported to be effective in treating hyperhidrosis in patients (Peterson *et al.*, 1982; Holzle and Ruzicka, 1986). Iontophoresis of just tap water has been reported to be effective for hyperhidrosis of palms, soles and axillae (Shrivastava and Singh, 1977; Akins *et al.*, 1987; Elgart and Fuchs, 1987). The addition of some chemicals to improve the effectiveness of tap water iontophoresis has also been investigated (Grice *et al.*, 1972; Abell and Morgan, 1974). The Drionic device has a direct current output of 0 to 10 mA, but the current reverses its polarity every 2 min, and is described by the manufacturer as alternating current. The advantage of alternating current is to decrease the accumulation of unwanted ions under the electrodes, which would otherwise result in pH changes during chronic use of the iontophoresis device. The Lectro Patch® discussed earlier also uses an alternating current. The use of this patch to deliver hydroxocobalamin has been claimed to avoid electrochemical burns (Howard *et al.*, 1995). A relatively recent study investigated the efficacy of AC/DC iontophoresis for hyperhidrosis and reported it to be equal to that of conventional DC treatment. However, no effect was seen when pure AC was used (Reinauer *et al.*, 1993). Other devices have also been investigated for treatment of hyperhidrosis (Midtgaard, 1986).

4.2.6 Diagnosis of cystic fibrosis

Iontophoresis devices also have a use in the diagnosis of cystic fibrosis, and the historical origin of this use is discussed in Section 1.4.2. Iontophoresis devices are prescription devices approved for the diagnosis of cystic fibrosis by iontophoresis of pilocarpine. Iontophoresis of pilocarpine for about 15 min results in profuse sweating which continues for about 30 min following the treatment. Sweat is collected during this period and analyzed for chloride. Several devices for collecting a fixed volume of sweat based on a colour change or other means have been patented (Fogt et al., 1984, 1989; Lattin and Spevak, 1984). The finding of a high sweat chloride value (above 60 mEq/l) on at least two occasions along with the presence of clinical features of cystic fibrosis is consistent with the diagnosis of the disease. In a typical protocol, the electrodes are removed following iontophoresis of pilocarpine, the skin site is thoroughly cleaned and a preweighed filter paper is placed on the skin area which was exposed to pilocarpine and sealed with a plastic wrap. Following the sweating period, the paper is removed, weighed immediately and used for sweat analysis.

A cystic fibrosis indicator (CF Indicator®, Scandipharm, Birmingham, AL) is commercially available (originally, a product of Medtronic, Inc.). The test typically takes 25 min, and changes in the outer brown ring of the test patch are noted. If there is no colour change, cystic fibrosis is not indicated, while a complete colour change would result if cystic fibrosis is indicated. A partial colour change requires retesting at a cystic fibrosis centre. The system is based on a paper patch which collects sweat and shows colour change when chloride levels exceed 40 to 50 mEq/l. In a study of 66 patients with cystic fibrosis, there were no false negatives (Yeung et al., 1984). Another study examined the relation between the chloride concentration and the complexed chloride of the CF Indicator® using a digitizer and computer. It was found that the sweat chloride concentration can be calculated with a reproducibility equal to that of the Gibson–Cooke sweat test (Warwick et al., 1990). The use of this cystic fibrosis indicator has been evaluated in a five-centre study. This device can deliver pilocarpine iontophoretically and has a disposable chloride sensor patch which absorbs a specified volume of sweat. The device was found to be comparable with other tests and was suggested to be potentially useful for physicians' offices, in clinics and similar settings (Warwick et al., 1986).

Another commercially available device which is approved by the Cystic Fibrosis Foundation is available from Wescor, Inc. (Logan, UT, USA). It consists of a Webster sweat inducer with Pilogel® discs for iontophoresis of pilocarpine (Figure 4.2). By activating a start switch, an optimal quantity of pilocarpine is delivered for gland stimulation, equivalent to 5 min of iontophoresis at 1.5 mA. The Pilogel® iontophoresis discs are hydrophilic hydrated gels containing pilocarpine (Webster, 1983). Sweat is collected on a Macroduct® collector (Webster and Barlow, 1985) and can be analyzed with a Sweat-Chek™ analyzer. The sweat flows between the skin and the concave undersurface of the Macroduct collector, and a water-soluble blue dye on the collection surface allows visualization of the accumulated volume. A typical volume collected is about 60 µl during a 30-min collection. The analyzer is based on conductivity measurements, which have been suggested to be as reliable as analysis for chloride. It has been recommended that hospital departments which perform sweat tests and paediatricians should inform their patient about the small risk of minor burns during iontophoresis of pilocarpine. However, with the improvements in electrode design, chemical burns should now be very infrequent (Rattenbury and Worthy, 1996). Optimization of factors affecting iontophoretic delivery of pilocarpine has been attempted using response surface methodology (Huang et al., 1995). Any potential role of insulin in cystic fibrosis has also been investigated. Iontophoresis of insulin was observed to cause a decrease in sweat chloride concentration (Shapiro et al., 1975).

Clinical applications of iontophoresis devices for topical dermatological delivery 65

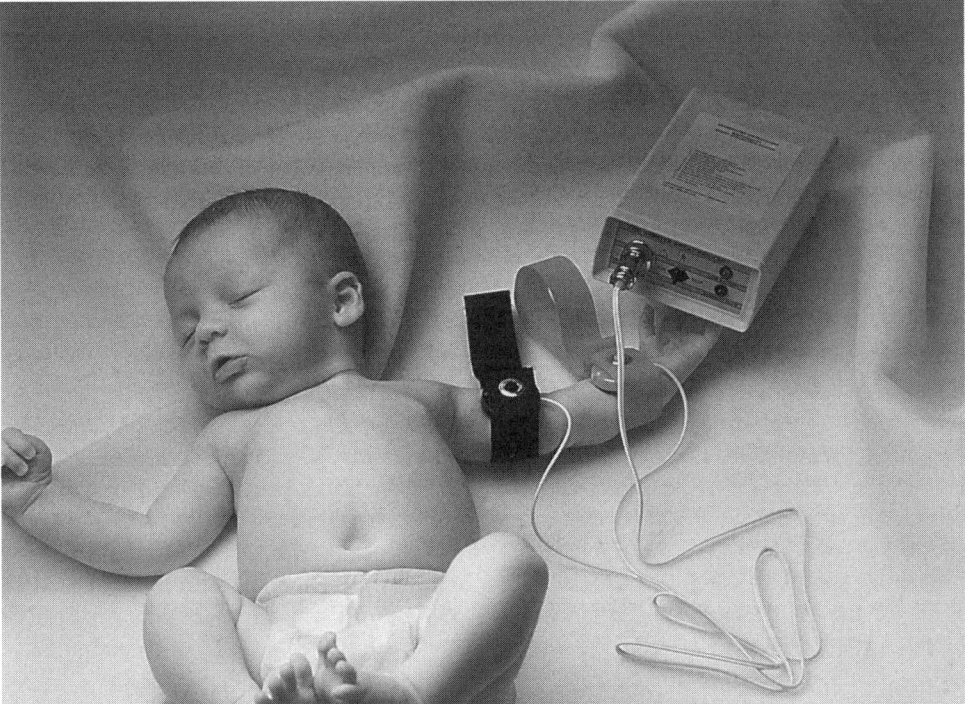

Figure 4.2 The Webster Sweat Inducer. Sweat will subsequently be collected by the Macroduct® sweat collector and analyzed by the Sweat-Chek™ analyzer. (By courtesy of Wescor, Inc., Logan, UT, USA.)

4.2.7 Iontophoresis of histamine and antihistamines

Skin weals induced by intradermal injection of histamine cannot be completely inhibited by H_1 blockade. The prick technique is less traumatic and leads to the formation of a more specific H_1-dependent weals. It has been suggested that iontophoresis of histamine can be even less traumatic and a useful replacement for intradermal injections or prick tests to evaluate the activity of antihistamine drugs in the skin by inhibiting the dermal reaction induced by histamine. Following iontophoresis (1.4 mA/cm² for 30 s) of histamine to humans, flares and weals developed rapidly and remained for at least 3 h. Thus, it seems that histamine iontophoresis can be a valuable tool in assessing the activity of H_1 blocking agents (Thysman *et al.*, 1995a). It has been reported that 4 h after administration of the potent H_1 blocker cetirizine, there was a 100 per cent inhibition of the weal in 9 of 10 human volunteers. This suggests that histamine iontophoresis could be the best method for inducing a 100 per cent H_1-dependent weal (Van Neste *et al.*, 1996). However, if an itch model with sustained itch half-life is desired, then the skin prick test was shown to produce a stronger and longer-lasting reproducible itch sensation than iontophoretic application. During the brief iontophoresis (10-s) period, histamine apparently passed the most superficial pruritoceptive C fibres too quickly to induce long-lasting itch sensations. On the other hand, some histamine was deposited at the dermal–epidermal junction when delivered by pricking, from where it was released throughout the time measurements (Darsow *et al.*, 1996). A pencil-shaped device for the local iontophoretic delivery of the antihistamine diphenhydramine hydrochloride has been investigated in six subjects. The pencil-shaped aluminium cartridge served as the cathode and had an external diameter of 14 mm and length of 92 mm. The system also had an electronic display indicating when the system was in use. At its lower end, it had a plastic isolator with an agar gel containing the drug, which served as the anode. The system was meant for local application of antihistamines to the skin, such as for the short-term treatment of acute skin irritations caused by insect bites. Release of the drug from the system on short-term use (5 min) was demonstrated, followed by a study in six subjects showing that a closed circuit is formed with an electric field in which drug transport can occur. The mean current flow was 86 µA and the mean electrical resistance of the body was measured at 116 kΩ (Groning, 1987).

4.2.8 Delivery of antiviral agents

Potentially, iontophoresis can be used for delivery of antiviral agents. The delivery of three antiviral agents, iododeoxyuridine, adenine arabinoside monophosphate and phosphonoacetic acid, was demonstrated several years ago in mice (Hill *et al.*, 1977). Another study showed that the antiviral drug idoxuridine (5-iodo-2′-deoxyuridine) can also be delivered iontophoretically to mice. Since both anodal and cathodal iontophoresis were found to assist its delivery through skin, it was inferred that electro-osmosis is an important factor in delivery (Gangarosa *et al.*, 1977). In a subsequent study on six subjects, iontophoresis of idoxuridine was conducted under the anode to 14 recurrent herpes labialis lesions. Immediate relief of discomfort and swelling and accelerated healing was reported, though the study was not blinded (Gangarosa *et al.*, 1979). Other studies have shown that iontophoresis can be a promising technique in delivering antiviral agents for benefit in herpes (Park *et al.*, 1978; Kwon *et al.*, 1979; Boxhall and Frost, 1984; Henley-Cohn and Hausfeld, 1984). Acyclovir (MW 225), a synthetic analogue of 2′-deoxyguanosine, is another agent useful for treatment of cutaneous herpes virus infections. The absorption profile of acyclovir topically applied to rats has been estimated and skin permeability was found to be very low (Yamashita *et al.*, 1993). Iontophoresis has been reported to enhance the delivery of acyclovir through nude mouse skin by about 30 times compared with passive transport. Acyclovir is an ampholyte with two ionizable groups (pK_a 2.4 and 9.2). At pH 3.0, it was delivered primarily by electrorepulsion as 20 per cent of the drug was in protonated form, while at pH 7.4, it was primarily delivered by electro-osmosis with anodal iontophoresis as the drug was mostly unionized at this pH (Volpato *et al.*, 1995). A combination of enhancers and iontophoresis has also been found to be promising in increasing the delivery of acyclovir across nude mouse skin (Lashmar and Manger, 1994). Clinical research supports the iontophoretic application of antiviral agents in post-herpetic neuralgia (Layman *et al.*, 1986) and vinblastine has been delivered by iontophoresis for cutaneous Kaposi's sarcoma lesions (Smith *et al.*, 1992).

4.2.9 Delivery of antibiotic/anti-infective agents

Iontophoresis can be used to introduce a bacteriocidal agent into skin prior to taking patient's blood to reduce the effect of contaminants responsible for large numbers of false positives (Haynes, 1992) or to kill micro-organisms on the surface (Jass *et al.*, 1995; Woodson, 1995). Iontophoresis can also deliver antibiotics such as penicillin across a burn eschar into the underlying vascular tissues, with the drug levels being 200-fold higher than when no current is applied. Bacteriocidal concentrations can be achieved in areas which are considered to be major sites of the origin of infection (Rapperport et al., 1965). Similarly, iontophoresis of antibiotics has been used for treatment of burned ear, since systemic administration of antibiotics is not effective owing to the avascularity of the ear cartilage. In contrast, iontophoresis was able to deliver high doses of antibiotics to the avascular infected ear cartilage (La Forest and Confrancesco, 1978; Greminger et al., 1980). In another study, five rabbits with ear burns were treated with gentamicin iontophoresis in one ear and gentamicin-soaked gauze on the other. Analysis of the ear cartilage demonstrated a 20-fold increase in the levels of gentamicin in the iontophoresis-treated ear, in comparison with only low levels in the gauze-treated ears (Macaluso and Kennedy, 1989). In a retrospective study of patients admitted in one hospital over a 16-year period, it was found that the incidence of ear infection was virtually eliminated when managed by iontophoresis of antibiotics (Rigano et al., 1992). Bacterial infections can occur easily in topical wounds and iontophoresis could be useful to deliver antibiotics to prevent or treat such infections.

4.2.10 Applications in veterinary medicine

A topical route of delivery is attractive to the farming industry because it is less labour intensive to apply drugs to an animal's skin than to administer more conventional dosage forms such as drenches, injections and inoculations. These systems are less likely to cause trauma and tissue damage or to interfere with nutritional status. Several studies have shown that chemicals can be transported across sheep and cattle skins in therapeutic amounts for topical or systemic effect and these have been reviewed (Pitman and Rostas, 1981). The route of passage of drugs across the skin of farm animals is of interest and has been investigated (Jenkinson et al., 1986). Topical dosage forms used in veterinary medicine include dusting powders, suspensions, lotions, liniments and creams. Products such as dips, sprays, pour-on or spot-on formulations are usually meant for systemic absorption and may include a penetration enhancer (Walters and Roberts, 1993). Alternatively, iontophoresis can be used to enhance drug delivery in veterinary medicine. Also, in certain skin disorders in animals, the infecting organism resides in the pilosebaceous unit, from which it is not easily eradicated by topical medication. In these cases, iontophoresis could be especially useful because of the appendageal pathways of drug transport. Iontophoresis of methylene blue under positive electrode has been reported to eradicate heavy infections of *Demodex folliculorum* from canine skin, while iontophoresis of potassium iodide under negative electrode alleviated *Trichophyton verrucosum* infection of bovine skin (Jenkinson and Walton, 1974).

4.2.11 Other applications of iontophoresis

Tap water iontophoresis for treatment of hyperhidrosis has been described earlier. Tap water iontophoresis using a hydrogalvanic bath is also used to achieve analgesia and hyperaemia of the treated region (Berliner, 1997). Several other isolated examples of the potential uses of iontophoresis are discussed in this section. Iontophoresis has applications for the delivery of narcotic analgesics. Iontophoretic delivery of morphine for postoperative analgesia in patients who underwent total knee or hip replacement resulted in reduced need for the patient to take intravenous meperidine via the patient-controlled analgesic device. Also, levels of morphine observed in the plasma were high enough to provide early postoperative pain relief (Ashburn et al., 1992). A formulation of morphine citrate salts suitable for iontophoresis has been developed (Stephen et al., 1997). Another narcotic analgesic, fentanyl (see Section 2.8.4), has been successfully delivered to humans in clinically significant doses following iontophoresis for 2 h (Ashburn et al., 1995). Since hair follicles are a major pathway for conventional DC iontophoretic drug delivery, the method may be ideally suited for delivery of minoxidil to hair follicles to stimulate hair growth. Minoxidil is a basic compound with a pK_a of 2.62 and most likely ionizes at the pyridine oxide. At pH 5, it exists mostly as the mono-cation (Kamath and Gangarosa, 1995). However, as minoxidil itself has a low solubility in water and no net ionic charge, its cationic derivatives have been synthesized for iontophoretic delivery. Each of the cationic derivatives is synthesized by reacting the minoxidil parent compound with an organic or an inorganic acid (Poulos et al., 1995).

Similarly, iontophoresis may have potential use for treatment of acne, boils and other skin disorders which are characterized by closed, blocked appendages in the epidermis of the skin. Iontophoresis under negative polarity apparently drives the ions generated by electrolysis into the skin to disrupt the blockage and to establish drainage from the appendages (Stephen et al., 1990). Other isolated studies on clinical applications of iontophoresis include lithium for gouty arthritis (Kahn, 1982), zinc for ischaemic skin ulcers (Cornwall, 1981), acetic acid for calcium deposits (Kahn, 1977) and traumatic myositis ossificans (Wieder, 1992), cisplatin for basal and squamous cell carcinoma of the skin (Chang et al., 1993), vinca alkaloids for chronic pain syndrome (Csillik et al., 1982), and silver for chronic osteomyelitis (Satyanand et al., 1986). Iontophoresis has also found application in allergy testing. Patients allergic to grasses were tested iontophoretically with extracts of different grasses. The results demonstrated that the biologically active constituents of orchard, red top, sweet vernal and June grasses can be transported readily into the skin of the persons allergic to timothy grass. The skin reactions produced by iontophoresis, in general, are found to parallel the dermal responses observed in the usual skin tests (scratch and intradermal). Iontophoresis of chromium in guinea pigs has been investigated in order to test the role of the technique in allergy testing for chromium (Wahlberg, 1970). The iontophoretic application of potassium iodide to human knees has also been investigated. It was found that the iodide is taken up only when electric current is applied. About 10 per cent of the iodide applied was noted to penetrate the skin, while X-ray fluorescence scanning of the volunteer's thyroid gland showed that the average iodine content in the gland is increased by more than 30 per cent (Puttemans et al., 1982). Iodine iontophoresis has also been used for reducing scar tissue in a patient (Tannenbaum, 1980).

4.3 Iontophoresis devices

While wearable electric patches are mostly still in development for systemic delivery, external palm or Walkman-sized iontophoresis devices have been used for several decades for topical delivery. Generally, iontophoresis devices are available only on prescription and would be considered misbranded if labelled for lay use or over-the-counter use. In the United States, the FDA regards iontophoresis kits as Class III devices. However, they have not undergone the series of clinical trials of safety and efficacy because they were marketed prior to the passing of the medical devices amendments in 1976. Instead, they are cleared under the easier 501(k) process that applies to grandfathered Class III devices. In some cases, the FDA has allowed these devices to be regulated as Class II devices, such

Clinical applications of iontophoresis devices for topical dermatological delivery

Table 4.1 Iontophoresis devices and electrodes on United States market

Manufacturer	Iontophoresis device		Iontophoresis electrode		
	Brand	Maximum dosage (mA∗min)	Brand (No. sizes available)	Composition of drug reservoir	Buffer system
Empi, Inc. (St Paul, MN)	DUPEL[a]	160	EBIE (3)	Polyester fleece	Ion exchange resin
Henley Healthcare (Sugarland, TX)	Dynaphor	Manual	Iotrode (2)	Cotton/rayon	0.9% saline; 0.5% $KHPO_4^-$
			Medipad (1)	Hydrogel	None
Iomed (Salt Lake City, UT)	Phoresor II	80	Trans Q1 & 2 (2)	Hydrogel	$Ag^+/AgCl$
			Trans QE	Hydrogel/sponge	$Ag^+/AgCl$
Life Tech, Inc. (Houston, TX)	Iontophor PM/DX	150	Meditrode (6)	Cotton/rayon	0.9% saline; 0.5% $KHPO_4^-$
	Microphor	80			
General Medical Co.[b] (Los Angeles, CA)	Lectro Patch®	Infinite	Lectro Patch (1)	Polyester fleece	Not required
	Drionic®	Infinite	Drionic (2)[c]	Polyester fleece/wool	Not required
Wescor Inc. (Logan, UT)	Sweat-Chek™	7.5	Pilogel® discs	Gel reservoir	I.N.A.
Scandipharm (Birmingham, AL)	CF Indicator®	16	Built-in	Hydrogel	$Ag^+/AgCl$

[a] Two-channel device (all other devices are one-channel); [b] these devices use AC current (1 cycle/2 min; all other devices use DC current); [c] for underarm or hand/foot use. Abbreviations: $Ag^+/AgCl$, silver/silver chloride electrode; I.N.A., information not available. Modified and expanded from Panus and Banga, 1997.

as in the case of devices for diagnosis of cystic fibrosis and Iomed's Phoresor device for use with iontocaine preparation (Gwynne, 1997). The pharmacist will have a significant role to play in the use of iontophoretic devices and formulations (Baharloo, 1987; Allen, 1992). To date, seven manufacturers market iontophoretic devices in the United States (Table 4.1). The three largest companies in the United States are Iomed Inc. (Salt Lake City, UT), Empi, Inc. (Minneapolis, MN) and Life Tech, Inc. (Houston, TX). The annual market is about US$ 50 million for anaesthetic applications and US$ 20 million for physical therapy (Gwynne, 1997). These companies market external palm-sized devices which are hooked up to electrodes for topical delivery of drugs. These iontophoretic devices contain a power source and two electrodes. In addition to the drug electrode, a second electrode is applied to the patient's skin to complete the circuit. This second electrode is referred to as the ground, dispersive or indifferent electrode. The power source is a direct current generator, with a current output of 0 to 4 mA. These are designed as constant current devices, that is, the voltage adjusts based on the skin resistance of the patient so that the current stays constant at the desired setting. The dosage is expressed as follows:

dosage (mA∗min) = current (mA) × time (min)

The maximum dosage (mA∗min) delivered by these devices varies, with a maximum of 80 to 160 mA∗min between various manufacturers (Table 4.1). It should be noted that the use of mA∗min units to express dosage can be confusing (see Section 4.2.4).

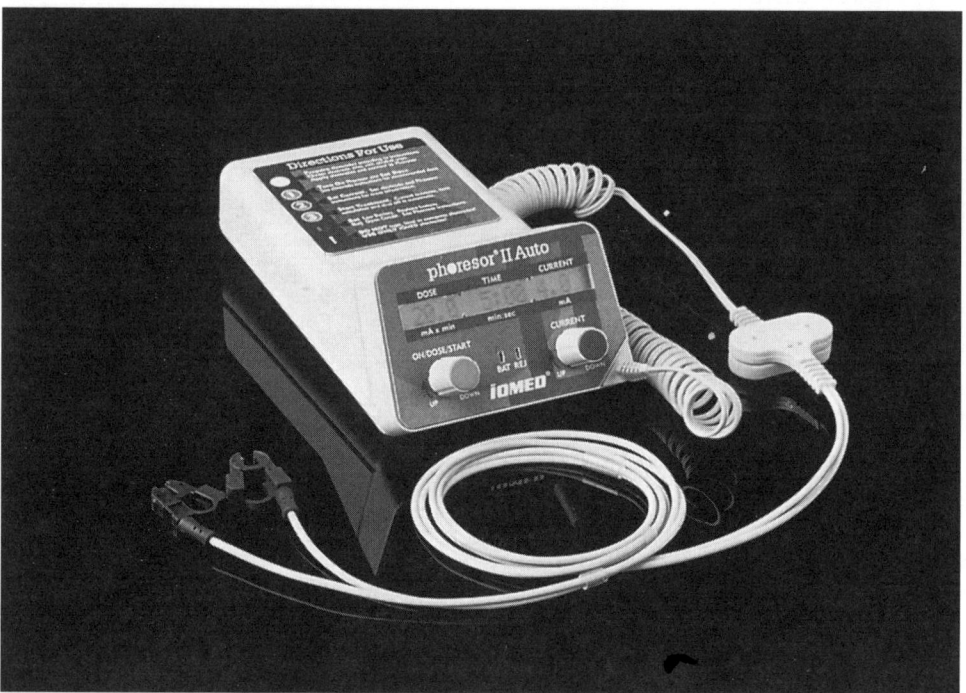

Figure 4.3 The Phoresor II system. (By courtesy of Iomed, Inc., Salt Lake City, UT.)

The Phoresor II Auto system is shown in Figure 4.3. It requires the setting of dose and current, by which it calculates the time required for the selected dose and adjusts it if current setting is changed during treatment. Current ramp-up and -down and automatic shut-off are provided. The DUPEL device from Empi is a dual-channel system, where the user can set the dosage and current levels for each channel independently. This can allow delivery at two different body sites at the same or different delivery rates, or two different drugs, or treatment of two patients simultaneously (Maurer *et al.*, 1993). The Iontophor PM/DX is a microprocessor-controlled programmable device, allowing the user to customize and store the most commonly used protocols. The devices from the General Medical Company are either wearable (Lectro Patch®) patches as discussed, or specialized for the treatment of hyperhidrosis (Drionic®), which were discussed under clinical applications of iontophoresis. The Lectro Patch® has iontophoretic circuitry with power to drive drugs through skin for 3 days at a high current setting or 8 days at a low current setting. For short-term use, this amounts to about 144 uses of 30 min each at high setting before the device must be replaced. The device can be used for multitherapy with two separate drugs, and for investigational studies by filling its drug reservoir pad with any specific drug. Since the current (AC) is periodically reversed at very low frequencies, there is no inactive electrode because both electrodes become active with alternating polarity and both contribute to driving the drug into the skin. As the polarity of each reservoir reverses regularly, the clinician need not be aware of the polarity of the drug to use this system effectively (Tapper, 1993).

Additional manufacturers of iontophoresis devices as found on the Internet may include Dagan Corporation (MN, USA), Generali Tecniche Elettroestetiche (Italy), Oriomedica (Italy), Deutsche Nemectron (Germany), Edith Serei Co., Ltd (Canada), and Ionto-Comed GmbH (Germany). Iontophoresis devices are also available for use in dentistry (Dentaphor®

II, Life Tech, Inc., Houston, TX). The device has both intra-oral and extra-oral applicators to apply drugs to relieve tooth hypersensitivity, administer local anaesthetics and alleviate suffering of temporomandibular joint disorder and myofacial pain syndrome. An iontophoretic toothbrush is also commercially available (hyG Toothbrush, Dyna Dental Systems, Phoenix, AZ, USA) and is claimed to remove the positively charged plaque from negatively charged tooth by attracting the plaque ions to the negatively charged bristles of the toothbrush. The toothbrush could also be used potentially for iontophoresis of fluoride in daily oral hygiene. Another electrical toothbrush (Ionoral®, Sintersan s.r.l., Italy) is claimed to transfer the fluoride from fluor-based toothpastes onto the dental porcelain in superior quantities compared with normal brushing. Iontophoresis devices are sometimes even used in beauty salons and figure salons such as the Dermaculture® facial clinics (Beverly Hills, CA, USA). In a similar application, Precis Dermalogics™ (Quorum International, Hong Kong) is a system which uses an iontophoresis applicator (Activator) and claims to deposit liposome-entrapped moisturizing and vitamin-rich ingredients into the layers of the skin.

4.4 Electrode design

Historically, iontophoretic devices were typically a crude assembly of paper towels, lint cloths, orthopaedic felt, gauze or other suitable means to apply current to the site of application (Stralka et al., 1996). These early electrodes did not provide uniform current density under the skin, resulting in the possibility of skin burns at sites which had a higher current density. Also, these electrodes did not provide any mechanism to prevent electrolysis of water or remove the extraneous ions being generated. The resulting pH changes could make the skin susceptible to irritation or burns and the mobile hydrogen and hydroxyl ions generated would lower the efficiency of drug delivery. To avoid these problems, electrodes are now carefully designed and many different types are commercially available. Electrode designs differ among manufacturers (Table 4.1). In general, the electrode design consists of a backing to which a carbon layer is added to distribute the electric current evenly over the drug delivery electrode. The drug reservoir consists of a hydrophilic layer to hold the drug solution during iontophoresis. The composition of the hydrophilic material varies between manufacturers, but typically consists of fibre-type materials or hydrogel polymer matrix. The fibre-type material varies from polyester fleece to rayon, and the natural fibres include wool and cotton. Several manufacturers utilize an additional outer layer which is in contact with the skin. This last layer serves to stabilize the underlying layers of the electrode and/or act as a wicking layer to absorb the applied drug solution. The exception to the above electrode design is Iomed, which utilizes a polymerized gel (hydrogel) as the drug reservoir. Many of the wearable, systemic delivery, iontophoresis devices which were discussed earlier as being in development also use hydrogels to hold the drug (see Section 3.5.2). A hydrogel formulation can provide a conductive base (if an appropriate solution is used), ease of application, uniform current distribution at the treatment site, and allow for replacement of a drug-loaded hydrogel pad, while reusing the device. Several hydrogel systems have been investigated for their use in iontophoretic delivery of peptides and proteins of different molecular weights (Banga and Chien, 1993a).

Electrochemical reactions taking place at the electrodes during iontophoresis could create several problems, as discussed in Section 3.4. The commercially available iontophoresis electrodes have used different mechanisms to avoid these changes (Table 4.1). The Trans Q electrodes (see Figure 4.4) from Iomed, Inc. use silver/silver chloride electrodes to

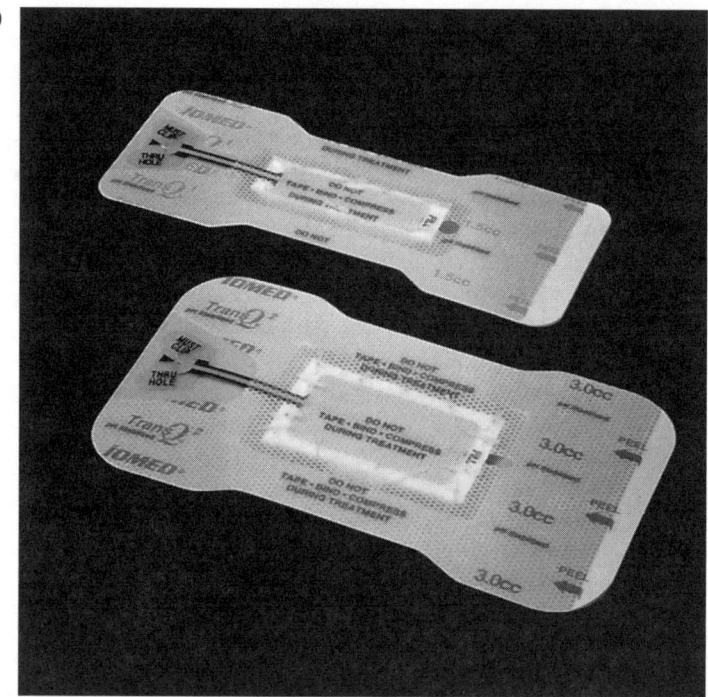

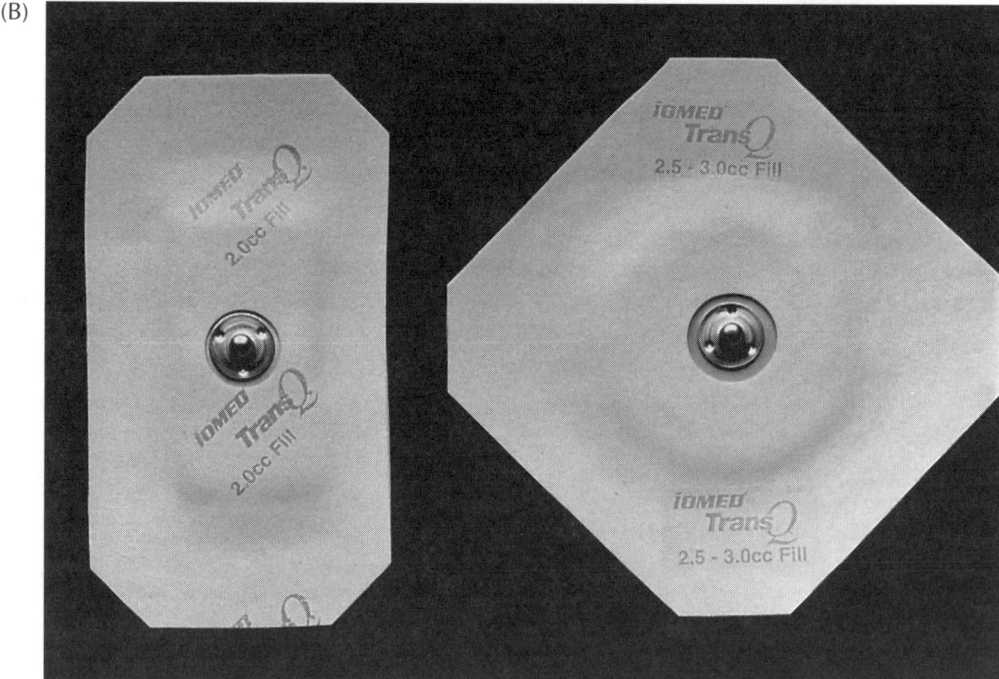

Figure 4.4 Trans Q electrodes. (A) Hydrogel-based Trans Q electrodes for use with the Phoresor II system. (B) Trans QE electrodes which are compatible with iontophoresis units from some other manufacturers. (By courtesy of Iomed, Inc., Salt Lake City, UT.)

stabilize drug pH without buffers. This becomes feasible because these electrodes actually participate in the electrochemistry, thereby preventing the electrolysis of water (see Section 3.4.2). Trans Q electrodes are available in two sizes, for larger or smaller treatment sites. Trans Q1 electrodes can be used with smaller sites such as fingers or toes. The larger Trans Q2 electrodes can be used on sites such as elbows, wrists, knees and shoulders. Trans QE electrodes are also available and are compatible with the devices of some other manufacturers, allowing the use of silver/silver chloride electrodes with devices other than Phoresor II. These various Trans Q electrodes are widely used in physical therapy, sports medicine and rehabilitation (Shaw, 1997). Meditrode electrodes (Table 4.1) are available in six different sizes spread over five different shapes to allow application over a variety of surfaces such as digits, elbows, ankles, shoulders and temporomandibular joints. The possible local silver toxicity via iontophoretic devices should be considered (Hollinger, 1996), but may be possible to avoid by appropriate electrode design.

Alternatively, buffers can be used to maintain the pH during iontophoresis. The buffering may be accomplished by the addition of a separate aqueous buffer such as phosphate to the electrolyte solution for drug delivery (Life Tech) or the impregnation of the phosphate into the electrode material (Henley Healthcare). However, even if buffers can maintain pH, hydrogen and hydroxyl ions being generated will compete with the drug for carrying the current, thereby decreasing the efficiency of delivery as a function of time (Sage, 1993). The buffer ions themselves will also provide extraneous competitive ions to the drug, resulting in lower drug delivery efficiency. Empi, Inc. includes an ion exchange resin in its electrode design, and the resin has been reported to stabilize pH under both cathodic and anodic conditions. The use of ion exchange resins in iontophoresis electrodes (Johnson and Lee, 1990; Petelenz et al., 1990) is discussed in Section 8.6. Finally, General Medical utilizes an alternating current at 0.008 Hz (1 cycle/2 min) to minimize changes in pH at the electrode (Table 4.1). As the current phase alternates at a regular frequency, the ions built up under the electrodes are much less. The low frequency AC current used generates hydrogen ions during one phase and hydroxide ions when the current reverses, thus apparently neutralizing pH changes and avoiding burns (Howard et al., 1995). The literature also describes a glass application chamber for human in vivo studies should there be a need to apply a liquid vehicle on the skin, without using any of the marketed devices with reservoirs containing absorbent material (Leopold and Lippold, 1992).

4.5 Treatment protocols and formulations

Previously published clinical protocols (Harris, 1982; Banta, 1994) and most practising clinicians apply iontophoretic treatments on an alternative-day treatment regimen. The number of treatments varies from clinician to clinician. Typically, patients are treated every other day, terminating after six to eight visits if significant relief is not obtained (Stralka et al., 1996). Setting up the unit takes about 5 min, after which constant supervision is usually not required. Typical treatment duration is 10–20 min using a current of 4 mA or less applied via an electrode such that the current density is less than 0.5 mA/cm^2. Prior to the application of the electrode, the skin is prepared by cleaning with isopropyl alcohol to remove any oily secretions or dead cells and also to disinfect. The site for the inactive electrode is similarly cleaned. The return or inactive electrode is used by all the devices listed in Table 4.1. The dispersive or inactive electrode should be placed about 15-cm from the active or drug electrode to prevent formation of edge currents from one electrode to the other. Any excessive hair may be clipped but not shaved. The electrodes are prepared according to the manufacturer's package instructions and should be applied over intact skin, not broken skin. If the electrode must be applied over any skin lesions, the lesion should first be protected with petroleum jelly. Iontophoresis should not be done over damaged or denuded skin or other recent scar tissue. The delivery electrode and the return electrode are applied in the appropriate anatomical locations. The iontophoretic device is connected such that the clinician's choice of electrodes, cathode or anode, is attached to the delivery electrode. The device in switched on and the delivery dosage or timer is set, depending upon the manufacturer of the iontophoretic device. Current density should be increased slowly and the patient should be informed that they may experience a tingling sensation initially. The current amperage is increased to the either a maximum of 4 mA or maximum tolerated by the patient. The iontophoresis continues until the dosage is delivered. The iontophoretic electrode is then removed by some clinicians; however, others leave the electrode in place to continue passive delivery of dexamethasone phosphate. After the iontophoresis treatment is over, a slight erythema (redness) under the electrode is normal and we found that it typically disappears in a few hours (Panus et al., 1997). The use of iontophoresis is contraindicated for patients with known adverse reactions or sensitivity to application of current or drug and those wearing cardiac pacemakers or other electrically sensitive devices. It is also contraindicated for treatment across the right and left temporal regions and for the treatment of the orbital region. The devices should obviously not be used if the patient is hypersensitive to the drug being used.

The major clinical use of iontophoresis is for the localized delivery of the anti-inflammatory agent, dexamethasone-21-phosphate. The pharmaceutic preparation of dexamethasone utilized is the same as that for injection. Dexamethasone-21-phosphate (Decadron® phosphate, 4 mg/ml) from Merck & Co. or a similar vendor is utilized. Some clinicians utilize dexamethasone phosphate (4 mg/ml) in combination with lidocaine (Xylocaine®, 10 mg/ml), according to the clinical protocols of Harris (1982) and Bertolucci (1982). The use of lidocaine in the iontophoretic solutions was proposed by Petelenz (Petelenz et al., 1992) to maintain pH. The dexamethasone solution is utilized either as formulated from the vendor, or diluted with an equal volume of sterile 0.9 per cent saline. No published source for the clinical iontophoretic use of this diluted drug solution exists. The iontophoretic dosage varies, but ranges from 40 to 80 mA*min at 4 mA or less. Both the dosage range and amperage vary depending upon the clinician, the anatomical site being treated, or the tolerance of the patient to the iontophoretic treatment. Unfortunately, no reference-based clinical standards have been established for dexamethasone iontophoretic dosage or amperage by either the physical therapy profession or the manufacturers of iontophoretic devices. Clinically, dexamethasone phosphate is delivered from the cathode or the anode, the latter electrode having been proposed to deliver dexamethasone by electro-osmosis (Petelenz et al., 1992). An appropriate sized electrode is selected. The dexamethasone phosphate solution is dispensed, usually 2.5 ml for the medium electrodes of most manufacturers, and applied to the delivery electrode. Iontophoresis is then conducted according to the treatment protocols described earlier. While formulations for iontophoresis are simple to make, some pharmacies (e.g. Vann Healthcare Services, KY, USA) will provide customized iontophoretic medications for patients for prescription products used by physical therapists.

5

Iontophoretic delivery of peptides

5.1 Introduction

In the coming years, therapeutic peptides and proteins are going to gain increasing importance as a result of rapid strides in the biotechnology industry. The therapeutic application and market introduction of this new generation of therapeutic agents will require parallel development of efficient delivery systems by the pharmaceutical industry. Because of their polypeptide nature, peptide and protein drugs are destroyed in the gastrointestinal tract and must be administered parenterally. These are invasive routes which involve the inconvenience of injections and add to the cost of the health care system if administered under medical supervision. Thus, non-invasive methods of administration would be preferred. The skin, with its accessibility, enormous surface area and possibility for site targeting, offers a potential means of non-invasive delivery. Several drugs are now available on the market as transdermal patches. However, none of these is a peptide or protein drug. This is because the skin is ordinarily permeable only to small lipophilic molecules, a criterion readily fulfilled by drugs such as nitroglycerine, scopolamine, clonidine, nicotine and other drugs on the market. Peptide and protein drugs, being hydrophilic and macromolecular in nature, do not readily permeate the skin.

The transdermal route offers some distinct advantages for the delivery of peptide drugs, in addition to the general advantages of transdermal delivery discussed earlier (Section 1.1). Since these drugs have short half-lives, the greatest benefit would be the fact that the transdermal route provides a continuous mode of administration, somewhat similar to that provided by an intravenous infusion. Transdermal delivery of peptides may become feasible if assisted by iontophoresis or other enhancement means such as electroporation, phonophoresis or chemical enhancers. Iontophoretic enhancement provides further benefit for peptide delivery because, theoretically speaking, the rate of peptide delivery can be initiated, terminated or accurately controlled/modulated merely by switching the current on and off or adjusting the current application parameters, respectively (Chien *et al.*, 1989; Parasrampuria and Parasrampuria, 1991; Cullander and Guy, 1992; Banga, 1996). This would be especially useful because a pulsatile delivery may be required for some peptides, as opposed to constant delivery. Also, the skin is relatively low in proteolytic activity, compared with other mucosal routes, thereby reducing degradation at the site of administration. It is important to have a sensitive, stability-indicating, validated assay to characterize the permeation profiles of the drug. Since proteins are very potent drugs, only small quantities are used and the amount of peptide permeating through the skin into the receptor in an *in vitro* set-up is generally too low to be within the sensitivity of HPLC assay with a regular UV detector. Radiolabelled peptides are thus commonly used, but the use of radiotracers can lead to erroneous results as the assay may be measuring proteolytic fragments rather than the intact protein. The use of a radiochemical detector on an HPLC may help to solve this problem. Similarly, once spread in the volume of distribution, the plasma concentrations during *in vivo* studies may be below the assay limits for some analytical methods. Analytical techniques such as radioimmunoassay and enzyme-linked immunosorbent assay may be used to characterize the plasma levels of peptides following transdermal delivery.

Iontophoresis has been widely used for topical delivery of conventional drugs (see Chapter 4). Similarly, it can be used for topical delivery of peptides if any local indications exist. For instance, the delivery of anti-infective peptides for local effect may be feasible. However, perhaps the most common use of therapeutic proteins for local effect on skin would be the area of wound healing. The potential use of iontophoresis in this field remains to be seen since iontophoresis is normally used on intact skin, not broken skin. Nevertheless, iontophoresis has shown promise for the delivery of lidocaine for wound healing (see Section 4.2.3). The most promising agent for topical dermatological use is the epidermal growth factor (EGF), a polypeptide of 53 amino acid residues, which has shown promise in healing open and burn wounds (Celebi *et al.*, 1994). It has been found that the presence of protease inhibitors is required in the formulation to stabilize EGF at the wound site (Okumura *et al.*, 1990; Kiyohara *et al.*, 1993). Formulations of fibroblast growth factor have also been reported to accelerate wound healing in a diabetic mouse model (Matuszewska *et al.*, 1994; Okumura *et al.*, 1996). The topical application of interferon for controlling eruptions of herpes has also been reported (Weiner *et al.*, 1989; Ophir *et al.*, 1995). However, this chapter will mostly discuss only the systemic delivery of therapeutic peptides by iontophoresis. A multitude of factors needs to be carefully considered for the proper design of an iontophoretic transdermal drug delivery system for delivery of peptides. These include considerations of the charges on the peptide molecule and the skin in relation to the environmental pH. The presence of proteolytic enzymes in the skin and the binding of peptides to the skin to form a reservoir (or depot) are some of the other important considerations. Preformulation data should be generated for the formulation development of a dosage form or the design of a drug delivery system to achieve optimum stability and maximum bioavailability. Peptide/protein drugs are known to have a strong tendency to become adsorbed to a variety of surfaces and this must be carefully evaluated as it could lead to misleading interpretations of data. Certain additives, such as albumin, can be used to minimize such surface adsorption. Another potential problem is self-aggregation of peptide/protein molecules, especially under stirring. A careful choice of electrode, pH, buffer and ionic strength is required. The possibility of a reversible electrode such as silver/silver chloride reacting with a peptide to cause precipitation and/or discoloration should be considered. In such cases, alternative pH control mechanisms should be considered. The charge on the peptide molecule can be controlled by adjusting the solution pH and thus delivery can be manipulated for either cathodal or anodal iontophoresis. The use of hydrogels could provide a pragmatic choice for a delivery system (see Section 3.5.2).

5.2 Structural considerations of polypeptides relevant to iontophoretic delivery

A basic understanding of peptide structure is required to appreciate fully the complexities involved in their electrically assisted transport into or across skin. The 20 different naturally occurring amino acids are the building blocks of peptides and proteins. All amino acids (except glycine) have a chiral carbon and are zwitterions, both in solution and in the solid state. The ionizable sites in the peptide such as the imidazolyl nitrogen of histidine, phenol hydroxyl of tyrosine or guanidine nitrogen of arginine should be known. Each functional group has a characteristic dissociation constant, K, whose negative logarithm is called the pK_a. Since pK_a values were measured at fixed ionic strength, they are sometimes called apparent dissociation constants, pK'_a. The constant pK'_1 usually refers to the most acidic group. For example, the dissociation constants for glycine, $pK'_1(COOH)$ and $pK'_2(NH_3^+)$ are 2.34 and 9.60, respectively. Aspartic and glutamic acid have a negative charge while lysine and arginine have a positive charge at the physiological pH of 7.4. The imidazole group in histidine carries a partial positive charge at pH 7.4. Serine and threonine have side chains that carry no charge at any pH, but are polar in nature. In contrast, tryptophan, phenylalanine and isoleucine have side chains which are more hydrocarbon-like in character. The pH at which all the molecules are in the zwitterionic form so that there is no charge on the molecule is called the isoelectric point (pI) of the amino acid. Of the 20 amino acids, 15 have isoelectric point (pI) near 6.0. The three basic amino acids have higher pI while the two acidic ones have lower pI (Banga, 1996).

Amino acids join with each other by peptide bonds to form polymers referred to as peptides or proteins. The distinction between peptides and proteins is somewhat arbitrary. Typically, peptides contain fewer than 20 amino acids and have a molecular weight less than 5000, while proteins contain 50 or more amino acids. Between these two categories are polypeptides which contain about 20–50 amino acids. Therapeutic proteins are generally referred to as globular proteins as they are of nearly spherical shape in solution. A protein has several levels of structure. These are generally referred to as the primary, secondary, tertiary and quaternary structures. The sequence of covalently bonded amino acids in the polypeptide chain is known as the primary structure and is determined genetically by the sequence of nucleotides in DNA. Once a polypeptide chain is formed, there is only one residue with a free amino group (N-terminal) and only one with a free carboxyl group (C-terminal). Peptide and protein chains are always written with the N-terminal residue on the left. The turns and loops of the polypeptide chain constitute the secondary structure. α-Helices and β-sheets are the common conformations of the secondary structure, with the rest being tight turns, small loops and random coils. In the α-helix, a single protein chain twists like a coiled spring. Each turn contains 3.6 amino acid residues, with the side chains pointing out from the helix. In the β-sheet conformation, the protein backbone is nearly fully extended. β-Sheets can be either parallel or antiparallel. In the parallel β-sheet, two or more peptide chains align their backbones in the same general direction. The antiparallel β-sheet has adjacent chains having parallel but opposite orientations. A fully extended polypeptide chain of about 60 residues may be about 200 Å long, but the folded globular protein may be just 30 Å in diameter.

The tertiary structure of a protein is the overall packing in space of the various elements of secondary structure. Protein folds in a highly specific fashion and it is believed that this folding is determined by the amino acid sequence or the primary structure of the protein. However, the folding is guided by some general rules: the overall shape is spherical, polar groups are on the surface while hydrophobic groups are buried in the interior, and the interior is closely packed. Guided by these rules, the protein folds into a unique structure. The forces that may stabilize the folded structure of a protein may be covalent or non-covalent. Disulphide bridges between cysteine residues provide a covalent linkage which can hold together two chains or two parts of the same chain. The non-covalent forces which can stabilize the protein include hydrogen bonding, salt bridges and hydrophobic interactions between the side chains of amino acid residues. Unlike proteins, peptides may display several conformations in solution. This is because they lack the hydrogen bonds and disulphide bridges that stabilize the three-dimensional structure of proteins.

5.3 Peptides as iontophoresis candidates

Analogous to amino acids, the isoelectric point (pI) of a protein occurs at the pH where the positive and negative charges are balanced. At their isoelectric point, proteins behave like zwitterions. At any pH below its pI, the protein has a positive charge and should be delivered under anode, and at pH above its pI, it has a negative charge and may be delivered under cathode. If a protein has an equal number of acidic and basic groups, its pI is close to 7.0. If it has more acidic than basic groups, the pI is low, and vice versa. For peptide drugs, the pH of the buffer used relative to the isoelectric point of the drug will determine whether the drug will have a positive or negative charge. The pH of the buffer should be at least one and preferably two pH units away from the pI of the peptide to ensure a high charge density on the peptide. Peptides with a high isoelectric point such as vasopressin (pI ~ 10.9) or salmon calcitonin (pI ~ 10.4) are perhaps the ideal candidates for iontophoretic delivery. This is because they will have a positive charge with high charge density at physiological or lower pH values. Since these will be delivered under anode, electro-osmotic flow which moves from anode to cathode (Section 2.6) will assist delivery in addition to direct electrostatic repulsion. In contrast, for peptides with very low pI (< 3) delivered from a pH 7.4 buffer, the electro-osmotic flow may hinder their transport. Thus, it will have to be established if optimal delivery takes place under the cathode or anode depending on whether direct electric repulsion or electro-osmosis is the predominant mechanism of transport. If the formulation has a low pH (< 4), the charge on the skin may be neutralized and a diminished or even reversed electro-osmotic flow is possible (Section 2.6). Iontophoretic delivery of peptides with a pI between 4 and 7.3 is faced with severe challenges (Sage *et al.*, 1995). In order for a peptide to be delivered iontophoretically through the skin, it should ideally not encounter any environment where its charge may reverse. The pH of the stratum corneum can range from 4 to 6, with a lowest pH of 3.5 to 4.5 at some small distance below the surface of the stratum corneum. The pH of the hydrated tissue just under the basement membrane is about 7.3. Thus, a peptide with a pI of, say 5, will be delivered into the skin up to some distance before it encounters a pH equal to its pI. At this point, the peptide will become uncharged and the iontophoretic force will no longer apply. As it diffuses little further because of the concentration gradient (or electro-osmotic flow), it will reverse its charge and may be pulled back towards the delivery electrode until it is again uncharged. Thus, the peptide will concentrate at some depth below the skin to form a depot or drug reservoir and may even precipitate in the skin (Sage *et al.*, 1995).

The pH of the receiver fluid can also be an important variable for *in vitro* studies since it will affect the pH of the viable epidermis (Kou *et al.*, 1993). In order for the study to represent physiological conditions, a physiological receiver fluid such as pH 7.4 phosphate buffer should be used. It has also been suggested that the electrically assisted transdermal delivery can be increased by non-ionic and zwitterionic surfactants for peptides and proteins which have at least one hydrophobic site. Apparently, the hydrophobic site impedes transport owing to interactions with the lipophilic region of the skin. The surfactant helps to shield the hydrophobic site, thus minimizing these interactions. It was shown that while cytochrome C, lysozyme and ribonuclease have similar molecular weight, their *in vitro* skin flux was different and correlated with their hydrophobicity (Holladay *et al.*, 1996). The salt form of the peptide can also be very important. The chemically enhanced *in vitro* permeability of leuprolide in the base form has been reported to be 10 times higher than in the acetate form due to improved lipophilicity of the base form (Lu *et al.*, 1992). For iontophoretic delivery, salt forms are more likely to be permeable than the base. Indeed, if the base is not water soluble, then salt forms may be the only choice. However, soluble peptides may also be available only in salt form owing to procedures used in bulk synthesis. Selection of proper salt form is important. If the hydrochloride salt is available and silver/silver chloride electrodes are being used, then it will at least partly obviate the necessity of adding chloride to drive the electrochemistry, thus improving efficiency of delivery by reducing the competitive ions. Since a peptide in aqueous solution is unlikely to have significant passive permeation, any comparison of the advantage of iontophoresis of aqueous solution of a salt of a lipophilic peptide should be with passive permeation of the base from a non-aqueous solution, rather than the aqueous salt formulation used in iontophoresis. The same is true for conventional drugs which may have been solubilized in aqueous solution for iontophoretic delivery.

The technique of capillary electrophoresis, which is normally used as a analytical tool, can also be used to screen promising peptide candidates as well as to optimize the formulation and delivery conditions for iontophoresis. The technique has been used to estimate the electrophoretic mobility of thyrotrophin-releasing hormone as a function of pH and ionic strength. The resulting data compared well with reported literature values for iontophoretic flux of the molecule under similar formulation conditions (Huff *et al.*, 1995). With a similar rationale, the techniques of isoelectric focusing and capillary zone electrophoresis were used as a tool to predict the ability of a peptide to be iontophoresed. Based on such studies, native luteinizing hormone-releasing hormone (LHRH) was predicted to be better suited for iontophoretic delivery than its free-acid analogue. The pI of LHRH was determined to be 9.6 while that of the analogue was 6.9. Furthermore, native LHRH was more mobile at pH > 2.5, though the two compounds were chemically similar (Heit *et al.*, 1994a).

5.4 Protease inhibitors and permeation enhancers for iontophoretic delivery

5.4.1 *Proteolytic activity of skin*

Though the skin is low in proteolytic activity compared with mucosal routes, it still retains substantial proteolytic activity to present a major barrier to delivery of peptides and proteins. The proteolytic enzymes in skin have been reviewed (Hopsu-Havu *et al.*, 1977; Steinstrasser and Merkle, 1995). Of the exopeptidases in the skin, aminopeptidases are the best known, while carboxypeptidases are in relatively smaller quantity or absent. The endopeptidases include the proteinases such as the caseinolytic enzymes, chymotrypsin- and trypsin-substrate hydrolyzing enzymes, thiol proteinase and carboxyl proteinases. In addition, proteinases attacking specific substrates such as collagenases, elastase, fibrinolysins, and enzymes of the kallikrein–kinin system are present. Most of these enzymes have not yet been isolated, purified or characterized. It must be realized that studies with skin homogenates may not be a good predictor of the proteolytic degradation actually encountered during transport. This is because if proteolysis in skin is discussed based on studies with skin homogenates, the structural complexity of the skin is ignored. The subcellular compartmentalization of the proteolytic enzymes may be such that these enzymes do not encounter the peptide or protein drug as the pathways of penetration may not pass through the cells that hold these enzymes. Once a skin homogenate has been made, the origins of the individual enzymes in the homogenate cannot be determined. Furthermore, the harsh techniques required to disrupt the cells during homogenization may destroy some enzymes. The enzymatic activity also differs between the stratum corneum side and the dermal side. When aqueous LHRH solutions were exposed to the stratum corneum side on excised hairless mouse skin, they were stable over a period of 40 h, but solutions exposed to the dermal side degraded by about 43 per cent after 40 h (Miller et al., 1990). The degradation may also be pH dependent. LHRH was stable when exposed to skin at pH 3, but was degraded at pH 5 and 7 (Chen and Chien, 1994).

The hair follicles and sebaceous glands may have higher metabolic activity than keratinocytes. Since the distribution of these appendages varies with anatomical site, this could be one contributing factor to the enzymatic activity being different at various anatomical locations. Another factor could be the varying ratio of epidermis to dermis at different sites (Steinstrasser and Merkle, 1995). The degradation of bovine serum albumin (BSA) during iontophoresis experiments with hairless mouse skin has been investigated by distribution of radioactivity in selected molecular weight ranges. While little change was seen in monomeric BSA (70–50 kDa) after contact with the stratum corneum or dermis for 24 h, it was converted to lower molecular weight species (15–1.5 and < 1.5 kDa) during transport (Figure 5.1) (Pikal and Shah, 1990b). The stratum corneum is known to have a stratum corneum chymotryptic enzyme which is a partially glycosylated 25-kDa molecule with an isoelectric point of about 10. It has an optimum pH of neutral to alkaline, but is also active at pH 5.5, the pH of the stratum corneum (Egelrud et al., 1996). Recently, a cell sheet model has been introduced as a tool to study epidermal metabolism and has been shown to have metabolic barrier properties comparable with human skin (Boderke et al., 1997). It should be noted that the proteolytic activity of skin from many animal species may be higher than that of human skin (Section 3.2.1).

5.4.2 Protease inhibitors

Protease inhibitors provide a viable means to circumvent the enzymatic barrier in achieving the delivery of peptides and proteins. Numerous agents could act as protease inhibitors by a variety of mechanisms, such as by tightly binding to or covalent modification of the active sites of proteases or by chelating the metal ions essential for proteolytic activity. The selection of an appropriate protease inhibitor could be guided by studying the principal proteases responsible for the degradation of the peptide/protein to be delivered, their subcellular compartmentalization and the mechanism for the transport of the peptide/protein. Protease inhibitors such as puromycin and amastatin were found to inhibit the degradation of leucine-enkephalin in skin homogenates of hairless mouse skin.

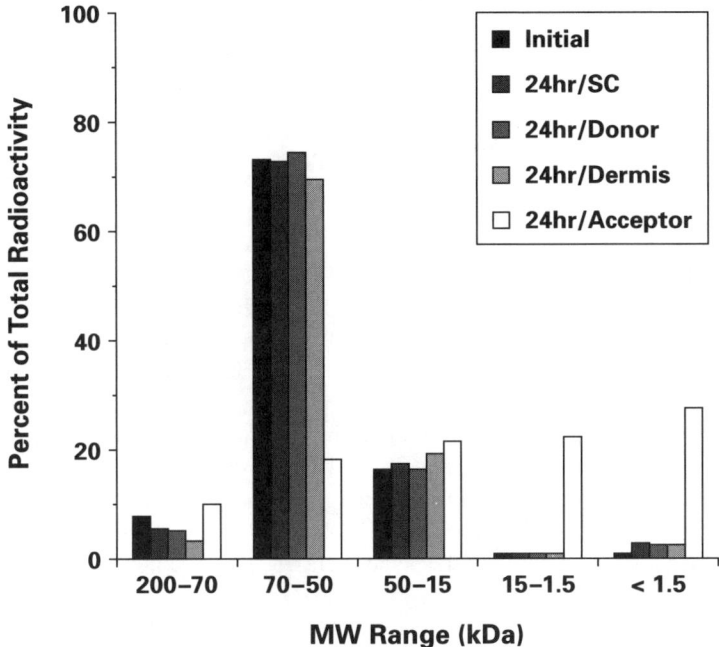

Figure 5.1 Decomposition of bovine serum albumin during iontophoresis experiments with hairless mouse skin at 37°C: distribution of radioactivity in selected molecular weight ranges. Distributions: Initial = prior to exposure to skin sample; 24 hr/SC = 24 h of exposure to stratum corneum without current flow; 24 hr/Donor = donor side (stratum corneum) after 24 h of cathodic iontophoresis at 0.32 mA/cm^2; 24 hr/Dermis = 24 h exposure to dermis without current flow; 24 hr/Acceptor = acceptor side (transported radioactivity) after 24 h of cathodic iontophoresis at 0.32 mA/cm^2. Mean equivalent solution flux, J_{vs}, based on total transported radioactivity is 1.2 ± 0.2 µl/h per cm^2. All solutions are 0.1 M NaCl and 0.01 M Tris buffer at pH 8.5 and contain 0.2 mg/ml non-radioactive bovine serum albumin. (Reprinted from *Pharmaceutical Research*, 7 Pikal and Shah, Transport mechanisms in iontophoresis. III. An experimental study of the contributions of electroosmotic flow and permeability change in transport of low and high molecular weight solutes, pp. 222–229, Copyright 1990 with permission from Plenum Publishing Corporation.)

Interestingly, these inhibitors were not very effective in skin diffusion experiments (Choi *et al.*, 1990). As discussed, this could be because the proteolytic enzymes may be present in membranes outside the actual diffusion path. Other potentially useful protease inhibitors include aprotinin, bestatin, boroleucine, *p*-chloromercuribenzoate, leupeptin, pepstatin, and phenylmethylsulphonyl fluoride. Sodium glycocholate, a penetration enhancer, may also act as a protease inhibitor. Protease inhibitors can be used to minimize the proteolytic degradation of peptides in skin during iontophoretic transport. This approach was found to be useful during delivery of calcitonin (see Section 5.5.5).

5.4.3 Permeation enhancers and iontophoresis

The combined use of permeation enhancers and iontophoresis has already been discussed (Section 1.5.3). In this section, specific examples for peptides are discussed. A non-ionic surfactant, n-decylmethyl sulphoxide, was found to increase the permeability of two amino acids, tyrosine and phenylalanine, through hairless mouse skin. This enhancer also increased the permeability of the dipeptide, Phe-Phe, and the pentapeptide, enkephalin (Choi et al., 1990). The percutaneous absorption of a tetrapeptide, hisetal (α-MSH), across hairless mouse skin was found to be enhanced about 28-fold when oleic acid was used as a penetration enhancer. While oleic acid was the most effective enhancer, azone and dodecyl N,N-dimethylamino acetate (DDAA), a relatively new penetration enhancer, were also found to be effective (Ruland et al., 1994a). Compared with hairless mouse skin, human skin was much less permeable to hisetal. Also, the effectiveness of penetration enhancers was much greater with mouse skin than human skin (Ruland et al., 1994b). The effect of enhancers on the permeation of vasopressin through excised rat skin is discussed in Section 5.5.3. The combined use of iontophoresis and penetration enhancers may allow greater amounts of drug to be delivered than either technique alone. In a study on transport of LHRH through porcine epidermis, iontophoresis was found to synergize with enhancers, such as 10 per cent oleic acid in combination with ethanol and 10 per cent oleic acid in combination with propylene glycol. Fourier-transform infrared (FT-IR) spectroscopic study showed that the combination of enhancer and iontophoresis increased the lipid fluidity, suggesting that the synergism in enhancement could be due to the greater fluidization of the stratum corneum lipids (Bhatia et al., 1997b). A somewhat similar observation was made for the *in vitro* transport of cholecystokinin-8 through porcine epidermis. Iontophoresis was found to increase the permeability of the drug further through enhancer-pretreated porcine epidermis than the control (Bhatia et al., 1997a).

5.5 Case studies

5.5.1 Amino acids

Iontophoresis of amino acids has been explored in an attempt to predict or understand the delivery of peptides. Amino acids may also have a direct benefit of moisturizing the skin as the presence of amino acids in the skin causes an increase in the penetration of aqueous solution into lipids (Coderch et al., 1994). Amino acids have also been used as model compounds to understand mechanisms of and factors affecting iontophoresis (Hirvonen et al., 1995). The passive permeability of amino acids from excised rat skin was found to vary with donor pH and amino acid, with the permeability coefficient being highest for di-cation, followed by mono-cation, positively charged, uncharged and negatively charged zwitterions (Hatanaka et al., 1996). For *in vitro* passive permeation through porcine skin, the lipophilicity of amino acids was found to be a more dominant factor than the molecular weight (Lin et al., 1996). It has been shown that the binding of a series of amino acids in excised abdominal skin of hairless rat decreased with an increase in the alkyl side chain (Wearley et al., 1990). This suggests that binding is likely to be polar or electrostatic in nature. We have investigated the iontophoretic delivery of amino acids by selecting one weakly charged amino acid, glycine (pI = 6.0), and one strongly charged amino acid, arginine (pI = 11.2). At the pH used (7.2), arginine would carry a strong positive charge while glycine would carry a relatively weak negative charge. As expected, arginine delivery was significantly more under the anode than under the cathode or passive conditions. For glycine, a higher delivery under the cathode was expected as it carries a negative charge. Instead, a higher delivery was found under the anode. The reason for this behaviour could be that electro-osmotic flow from the anode to the cathode overshadows the effect of polarity on the ionic molecule as the charge on the ion is relatively weak. Starting with a 5.0 mM concentration for both arginine and glycine, the delivery under the anode is about 23-fold higher for arginine compared with glycine. As discussed, this is because arginine, a very basic amino acid, carries a strong positive charge at the pH used, while glycine, a neutral amino acid, carries a relatively weak negative charge at the pH used. It is also possible that as glycine enters the skin, the pH of the environment is closer to 6.0, so that the molecule is essentially neutral in the solution (Banga et al., 1995b).

Another study delivered a series of amino acids across excised hairless mouse skin to investigate the effects of permeant charge (neutral, +1 or −1), lipophilicity and vehicle pH. Iontophoretic flux of zwitterions was significantly greater than their passive transport and delivery from the anode was greater than from the cathode, presumably due to electro-osmotic flow. For zwitterions (unlike the negatively charged amino acids), the iontophoretic/electro-osmotic flux did not reach steady state under the experimental conditions used and was inversely proportional to the permeant octanol/pH 7.4 buffer distribution coefficient (Green et al., 1991a). Amino acids and their derivatives with blocked amino or carboxyl groups have also been used to investigate the relative importance of the electro-osmotic flow in comparison with passive and electrical flow in transdermal iontophoresis across excised porcine skin. As expected, electro-osmotic flow enhanced anodic delivery of amino acids and flux decreased with increasing salt concentration in the buffer. However, the negative effect of electro-osmotic flow on anionic and neutral solutes was reduced as buffer concentration increased (Lin et al., 1997).

5.5.2 Small peptides

Thyrotrophin-releasing hormone (TRH) is a tripeptide with a molecular weight of 362 and a pK_a of 6.2. A prodrug approach has been used for transdermal delivery of TRH. Results of diffusion experiments using excised human skin indicate that the N-octyloxycarbonyl derivative showed an enhanced permeability. The prodrug that penetrated into the receptor phase was found to exist primarily as TRH (Moss and Bundgaard, 1990). Iontophoretic transport of TRH across excised, dorsal, nude mouse skin was found to be directly proportional to the applied current density. In the absence of current, the flux of TRH across skin was undetectable. However, even uncharged TRH was transported with iontophoresis, presumably by the electro-osmotic or convective flow that accompanies iontophoresis (Burnette and Marrero, 1986). Using response surface methodology to optimize the transdermal delivery of TRH, the maximum rate of permeation across rabbit inner pinna skin was achieved at low ionic strength, moderate pH and large current duty cycle under the constraint of a pulsed current density of less than 1 mA/cm^2 (Huang et al., 1996). In another study, flux of TRH across excised rabbit pinna skin in steady state was found to be proportional to the applied current density. At low ionic strength, the protonation of TRH was greater and its flux at pH 4 was greater than at pH 8. At higher ionic strength, the trend was reversed, presumably owing to the enhanced flux of unprotonated TRH due to electro-osmotic flow (Huang and Wu, 1996a). The stability of TRH under an electric field has also been investigated. Using platinum wires as electrodes, the degradation increased when current density was increased to more than 0.32 mA/cm^2, with rapid degradation at 0.64 mA/cm^2. After iontophoresis for 10 h, more than 75 per cent of the drug was degraded (Huang and Wu, 1996b).

Another tripeptide, threonine-lysine-proline (Thr-Lys-Pro), was successfully delivered across nude rat skin by iontophoresis under both *in vitro* and *in vivo* conditions. Transport of the tripeptide, which was positively charged at pH 7.4, was significantly enhanced by iontophoresis compared with passive transport. The delivery was found to be directly proportional to the applied current density over the range 0.18–0.36 mA/cm^2. Following 6 h of iontophoresis, 98.4 per cent of the radioactivity in the donor was still the intact peptide, while 94.0 per cent of the radioactivity that penetrated into the receptor phase was the parent Thr-Lys-Pro, suggesting that degradation in the skin was not significant (Green et al., 1992). The delivery of eight tripeptides of the general structure alanine-X-alanine, has been investigated across hairless mouse skin in vitro. The tripeptides exhibited very little degradation under the conditions of the experiment. The normalized iontophoretic flux was found to be independent of lipophilicity but inversely related to molecular weight. Steady-state fluxes were not achieved, suggesting that time-dependent changes in the properties of the skin barrier may be occurring (Green et al., 1991b). Iontophoretic delivery of another tripeptide, enalaprilat, has also been investigated across hairless guinea pig skin (Gupta et al., 1994b).

5.5.3 *Vasopressin*

Vasopressin, an antidiuretic hormone, and its analogues have been investigated for transdermal delivery. A nonapeptide with a molecular weight of 1084, vasopressin makes a good model peptide for investigation because of its ideal molecular size and isoelectric point as an iontophoresis candidate. Arginine vasopressin has a hexapeptide ring, in which two cysteine residues form a disulphide bridge, and a tripeptide tail. Owing to the presence of basic arginine and blocking of the C-terminus with NH_2, arginine vasopressin has an isoelectric point of about 10.9. Thus, more than 99 per cent of the drug is protonated in a buffer solution with a pH lower than 9.0. Vasopressin also has therapeutic use in the treatment of diabetes insipidus and emergency treatment of bleeding from oesophagogastric varices. Furthermore, vasopressin may require a pulsed delivery for therapeutic effect rather than steady-state levels. This is due to the possibility of tolerance or desensitization of receptors by its continual presence at a receptor site (Banga, 1996). Iontophoretic delivery can provide for such pulsed delivery by controlling the current. A brief overview of studies on transdermal delivery of vasopressin will be provided before discussing the studies with iontophoretic delivery. Effects of enhancers such as DMSO, sodium lauryl sulphate, azone and others on the transdermal permeation of vasopressin across excised rat skin have been investigated (Banerjee and Ritschel, 1989a, b). While sodium lauryl sulphate had no significant effect on permeation, azone was found to be the most effective enhancer and increased flux about 15 times. These findings were confirmed by in vivo studies on Brattleboro rats. These rats are genetically deficient in vasopressin and secrete large volumes of urine with low osmolality. Significant reduction in urine volume and increase in osmolality over a 24-h period resulted upon enhancement of vasopressin absorption in the presence of azone. In another study, the potential of azacycloheptan-2-ones (azones) as penetration enhancers for desglycinamide arginine vasopressin (DGAVP) was investigated. Using human stratum corneum under in vitro conditions, it was found that the hydrocarbon chain length of azone determined its effectiveness as a penetration enhancer. While pretreatment with hexyl- or octyl-azone did not enhance the flux, the permeability increased 1.9-, 3.5- or 2.5-fold after pretreatment with decyl-, dodecyl- or tetradecyl-azone, respectively (Hoogstraate et al., 1991).

Transdermal iontophoretic delivery of vasopressin across excised hairless rat skin has been investigated (Chien et al., 1990; Lelawongs et al., 1989, 1990). The delivery was found to be reversible upon termination of iontophoresis, with flux returning to that by passive diffusion alone (Lelawongs et al., 1989). It required about 1.5 to 2 h of iontophoresis treatment for vasopressin permeation to reach steady state (Lelawongs et al., 1990). Transdermal iontophoretic transport and degradation of vasopressin (spiked with labelled peptide) across human cadaver skin has been investigated by the author using a current density of 0.5 mA/cm^2 and analyzing results by an HPLC using a radiochromatography detector. Since a radiochemical detector was used for analysis, any free tritium or degradation products would not show up at the vasopressin retention time on the chromatogram. A higher degradation was observed in the receptor where the peptide was in contact with the dermal side of the skin. Higher enzymatic degradation in the receptor solution is most likely because more proteolytic enzymes would be present on the dermal side of the skin compared with the stratum corneum, which was in contact with the donor solution. This particular experiment was designed so that no transdermal transport took place during the experiment. Since vasopressin was mostly protonated at the pH (7.20) of HEPES buffer during the study, only anodal delivery was investigated. The cumulative amount of intact vasopressin permeated during 8 h of iontophoretic transport was 15.37 ($\pm$ 5.31; n = 3) µg/cm^2, which corresponds to only about 1 per cent permeation. Commercial products of vasopressin currently on the market include Pitressin® (Parke-Davis, Morris Plains, NJ), which contains 20 pressor units of vasopressin. One adult human dose is about 10 units and this corresponds to about 25 µg. This suggests that therapeutic doses can easily be administered using an iontophoretic patch with a realistic surface area, even though the cumulative amounts permeated correspond to only about 1 per cent permeation. Of the total radioactivity permeated, only about 40 per cent was intact vasopressin by 12 h. Several degradation peaks could be seen in the chromatogram. This suggests that the human cadaver skin still had enough enzymatic activity left to cause enzymatic cleavage of vasopressin into several products. The number of degradation products could be even higher than was apparent from such a chromatogram because any cleaved fragment that did not carry the tracer would not show up on the chromatogram. No intact vasopressin was found to permeate under passive conditions in this study. The flux of vasopressin across the skin was observed to stop as soon as the current was terminated, suggesting reversibility of delivery. Thus, modulation of peptide delivery by multiple applications of current should be possible. This can be used to achieve pulsatile delivery of peptides, which would be desirable as many physiological peptides are released in a pulsatile manner (Banga et al., 1995a).

A vasopressin analogue, 9-desglycinamide, 8-arginine vasopressin (DGAVP) is a neuropeptide drug which has been shown to improve the memory process in humans and is also used as a model peptide that is more potent and more resistant to metabolism than vasopressin. Iontophoretic delivery of DGAVP across human and snake skin has been compared (see Section 2.3). In another study, DGAVP was iontophoretically delivered through dermatomed human skin. The transport was observed to be controlled predominantly by electro-osmosis and the flux appeared to be controlled by the applied voltage rather than by the current density (Craane Van Hinsberg et al., 1994). Iontophoretic delivery of desmopressin acetate (DDAVP) has been investigated in vivo with a diabetes insipidus model in rats. Repeated short iontophoretic treatments with low current density were found to be best in maintaining a constant response. Using pulsed current, the antidiuretic response was dependent on current duty (on/off ratio) but not frequency (Nakakura et al., 1995, 1996). In another study, the prolongation of the antidiuretic response to desmopressin acetate in diabetes insipidus rats has been compared with other routes of administration. Delivery by iontophoresis was comparable with the nasal route when a dose about five times higher than the nasal route dose was used. In comparison with the oral route, iontophoresis was two to three times as effective (Nakakura et al., 1997).

5.5.4 LHRH and analogues

Luteinizing hormone-releasing hormone (LHRH), also known as gonadotrophin-releasing hormone (GnRH) is a naturally occurring decapeptide that stimulates the release of the pituitary gonadotrophins, LH and follicle-stimulating hormone (FSH). LHRH is secreted in a pulsatile manner and its replacement for treatment of primary infertility of hypothalamic origin requires pulsatile administration every 60 to 90 min through programmable pumps (Vickery, 1991). LHRH is readily degraded in the body by proteolytic enzymes and has a half-life of only 8 min. Several agonists and antagonists of LHRH have been synthesized and are promising agents for a range of clinical applications. Compared with LHRH, these agonists have relatively long half-lives and they can mimic the effects of continuous LHRH infusion if administered twice daily. Such administration stimulates the release of pituitary gonadotrophins initially, but repeated doses abolish the stimulatory effects. Owing to receptor desensitization and down-regulation, a stage of chemical castration results in about 4 weeks. Chronic application of LHRH agonists in controlled release formulations such as microspheres has proved effective in treating hormone-dependent cancers, prostate and breast cancer and in the treatment of sex-hormone-dependent disorders, endometriosis and uterine fibroids (Banga, 1996). Transdermal delivery provides a non-invasive route for the delivery of LHRH and its analogues. Chemical enhancers have been reported to enhance significantly the permeability of LHRH (Bhatia and Singh, 1997) and its analogues, leuprolide acetate through nude mouse skin (Lu et al., 1992), and nafarelin acetate through human cadaver skin (Roy and Degroot, 1994).

LHRH and an antagonist have been shown to be iontophoresed rapidly through hairless mouse skin, but metabolism by the skin to fragment peptides was observed (Miller et al., 1990). Degradation of LHRH by iontophoretic transport across porcine skin has also been reported (Bhatia et al., 1997b). The ability to deliver LHRH transdermally by iontophoresis has been demonstrated using the isolated perfused porcine skin flap model (IPPSF) (for details of model, see Section 3.3.3). A current of 0.2 mA/cm^2 was applied for 3 h and the mean LHRH delivered in 5 h using 21 skin flaps was 959 ± 444 ng. LHRH concentrations increased throughout the current application period and then decreased when the current was stopped. The flux data from IPPSFs were used as input into a two-compartment mamillary model and the predictions were checked by in vivo study in the pig. It was found that the model is able to predict the in vivo serum concentration of the iontophoretically delivered LHRH. Increases in FSH and LH concentrations were observed with increasing LHRH concentrations, indicating that iontophoretically delivered LHRH is biologically active (Heit et al., 1993). In another study using the IPPSF model, a drug depot was identified in the skin underlying the electrode which was approximately twice as large as the entire mass of drug delivered systemically. A small portion of this skin depot could be delivered with just a saline electrode during a second application period (Heit et al., 1994b).

Therapeutic doses of leuprolide, an LHRH analogue, have been delivered in 13 normal men using a double-blind, randomized cross-over study conducted under an IND granted by the FDA. The patches used were about 70 cm^2 and contained two reservoirs and an intrinsic power source supplied by small batteries. The electrical circuit was completed when the patch was applied to the skin. Passive patches were identical except that the wiring to complete the circuit was missing. Data analysis by ANOVA showed significant differences between the active and passive patches. The magnitude of elevation of LH produced by the active patches was in the therapeutic range and comparable with that achieved by subcutaneous administration. The only adverse effect reported was a mild erythema at the site of the active patch in 6 of the 13 subjects. This erythema resolved rapidly without sequelae (Meyer et al., 1988). Further evaluation compared the acute biological effect seen after subcutaneous and iontophoretic delivery of leuprolide to 18 volunteers. It was observed that the two techniques were very similar in the magnitude and duration of response achieved. The onset of the LH response was more rapid after subcutaneous administration since the period of absorption required for transdermal absorption was not a factor in this case (Meyer et al., 1990). A potentiometric titration curve of leuprolide shows that it has a net charge of +1 at a pH of 7.5 (protonation of Arg) and +2 at a pH of 4.5 (protonation of His). The enhancement factor for leuprolide across a synthetic membrane as a function of applied voltage at pH 4.5 was thus found to be double that at pH 7.5, directly proportional to the charge (Hoogstraate et al., 1994).

Nafarelin is another potent analogue of LHRH, which is commercially available as a nasal solution for the treatment of endometriosis and central precocious puberty. Transdermal delivery could provide an alternative route and use of chemical enhancers to facilitate the in vitro permeation of nafarelin acetate through human cadaver and monkey skin has been investigated. The flux from human cadaver skin from a propylene glycol/azone vehicle was 0.14 µg/cm^2 per hour. The steady-state lag time ranged from 24 to 40 h. Monkey skin was slightly more permeable than human cadaver skin (Roy and Degroot, 1994). Iontophoretic transport of nafarelin across hairless mouse skin (Delgadocharro and Guy, 1995; Delgadocharro et al., 1995) and human skin (Bayon and Guy, 1996) has also been investigated. Anodal delivery of positively charged nafarelin was found to result in a strong association of the lipophilic cation with the fixed negative charges on the skin, resulting in neutralization of the skin charge and reversal of electro-osmotic flow. Owing to this phenomenon, the peptide was found to inhibit its own delivery as a function of increasing concentration. An increased iontophoretic flux across hairless mouse skin was observed when the donor concentration was doubled from 0.5 to 1.0 mg/ml, but further elevation in concentration to 2 or 3 mg/ml did not increase the flux (Delgadocharro et al., 1995). In the case of human skin, the cumulative amount of nafarelin delivered iontophoretically (0.3 mA/cm^2) in 24 h actually decreased from 27.3 to 5.9 nmol/cm^2 as the donor concentration was increased from 0.1 to 1.0 mg/ml. The average delivery rate was about 1 µg/cm^2 per hour and will allow delivery of a target dose with a patch of 10–20 cm^2. Some metabolism of nafarelin during electrotransport was also observed (Bayon and Guy, 1996). The same has been seen with positively charged leuprolide (also see Section 2.6). However, the parent peptide (LHRH) does not impair electro-osmosis, leading investigators to search for a structural motif which may be responsible for this behaviour. It was observed that the inhibition of electro-osmosis by oligopeptides is reduced as the lipophilicity of the molecules is reduced. Also, the presence of a positive charge in close proximity to the hydrophobic portion of the molecule was considered essential. The carboxylate groups which give skin its negative charge probably interact with the positive charge of the oligopeptide while the adjacent lipophilic group on the peptide anchors the molecule somewhere along the pathway for iontophoresis. This changes the permselectivity of the skin and progressively reduces the electro-osmotic flow (Hirvonen et al., 1996).

Permeation of another analogue, buserelin, through isolated human stratum corneum by iontophoresis has also been reported. In this study, it was found that passive permeation of buserelin through human stratum corneum was not feasible. However, permeation was achieved and controlled by iontophoresis (Knoblauch and Moll, 1993).

5.5.5 Calcitonin

Calcitonin is a polypeptide hormone secreted by the parafollicular cells of the thyroid gland. It is a 32-amino acid polypeptide chain with a disulphide bridge and a molecular weight of about 3600. It is commercially available as calcitonin human and calcitonin salmon, and is used in the management of Paget's disease, hypercalcaemia and post-menopausal osteoporosis. Calcitonin salmon is more potent, with a dose which is about 50 times less than that of calcitonin human. Human calcitonin has a marked tendency to aggregate in aqueous solutions (Arvinte et al., 1993; Bauer et al., 1995; Cudd et al., 1995). Salmon calcitonin is more resistant to aggregation though it can also undergo base-catalyzed dimerization and sulphide disproportionation (Windisch et al., 1997). A nasal product has been commercialized recently (Physician Desk Reference, 1996). However, treatment with calcitonin is usually prolonged and thus repeated administrations would be required. Transdermal delivery could allow for continuous input. Electrical enhancement mechanisms, such as iontophoresis and/or electroporation, can modulate drug delivery by adjusting the current/voltage profiles so that pulsatile delivery is also feasible, when required. This continuous input could be used for chronic conditions such as Paget's disease and osteoporosis, while a bolus dose could be given for control of hypercalcaemic emergencies. Salmon calcitonin (pI = 10.4) and human calcitonin (pI ~ 9) have very high isoelectric points and so are good candidates for iontophoretic delivery.

The *in vivo* iontophoretic delivery of salmon calcitonin using male Wistar rats has been investigated. A hypocalcaemic effect was noticed when delivered under the anode from a pH 4.0 formulation. This was enhanced by the proteolytic enzyme inhibitors, aprotinin and camostat mesilate. No enhancement was seen with soybean trypsin inhibitor. This could be because aprotinin itself is a peptide (MW 6500, pI = 10.5) which is positively charged under most pH conditions and may itself be iontophoretically driven into the skin to provide protection. On the other hand, soybean trypsin inhibitor (MW 8000, pI ~ 4.1) was possibly neutral at the pH used and was not delivered into skin in significant amounts (Morimoto et al., 1992). Iontophoretic delivery of human calcitonin has also been investigated both *in vitro* and *in vivo* using hairless rats and a hypocalcaemic effect was observed (Thysman et al., 1994a). In an *in vivo* study with hairless guinea pigs, a significant increase in the transdermal electrotransport of salmon calcitonin was seen in the presence of a surfactant, polysorbate 20 (Holladay et al., 1996). A more recent study has delivered a calcitonin analogue to human volunteers by iontophoresis. A plasma concentration–time profile resulting from 6 h of iontophoresis was found to have a similar profile to intravenous infusion, but the appearance of the peptide in plasma following iontophoresis was slower than that during intravenous infusion (Figure 5.2) (Green, 1996b). Another recently published study found that transdermal iontophoresis of salmon calcitonin under the anode in rabbits caused a decrease in serum calcium levels, while no significant fall was measured in the absence of electric current. Substantial degradation of the polypeptide under the effect of electric current was observed. Owing to stability concerns in solution, a solid formulation was designed as a thin dry disc made by gentle compression

Iontophoretic delivery of peptides

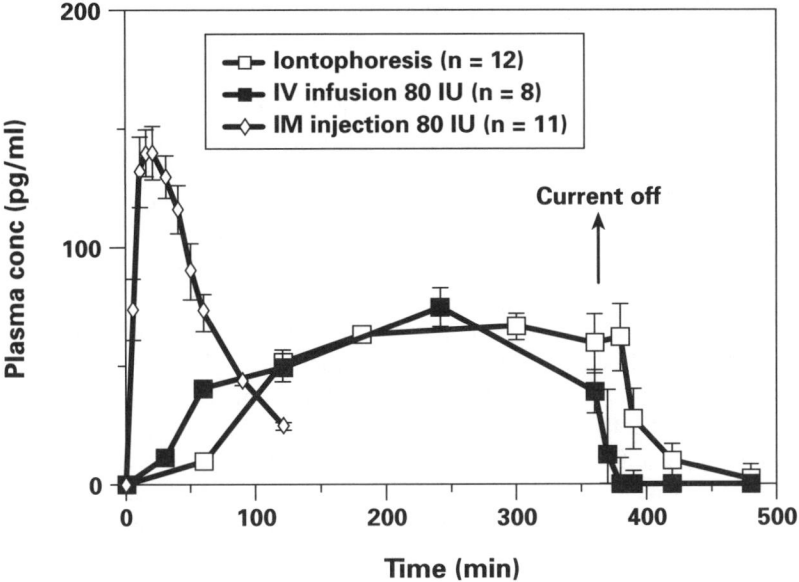

Figure 5.2 Plasma concentration–time profile resulting from a 6-h iontophoretic delivery of a calcitonin analogue peptide in healthy male volunteers. Plasma levels arising from intravenous (IV) and intramuscular (IM) injection of the peptide are also shown. □, iontophoresis ($n = 12$); ■, IV infusion 80 IU ($n = 8$); ◇, IM injection 80 IU ($n = 11$) (Reprinted from *Journal of Controlled Release*, **41** Green, Iontophoretic delivery of peptide drugs, pp. 33–48, Copyright 1996 with kind permission from Elsevier Science – NL, Sara Burgerhartstraat 25, 1055 KV Amsterdam, The Netherlands.)

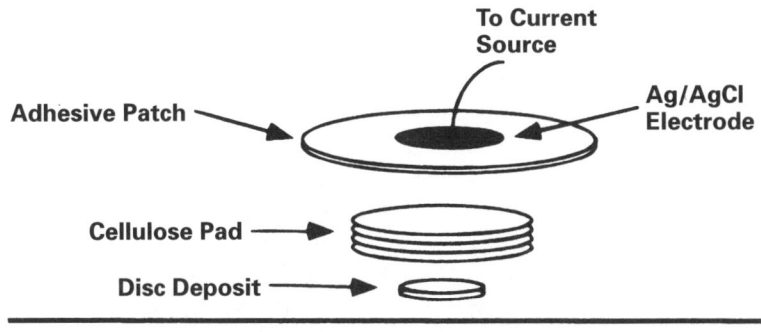

Figure 5.3 Sketch of the calcitonin iontophoretic delivery system containing the disc reservoir. (Reprinted from *Pharmaceutical Research*, **14** Santi *et al.*, Drug reservoir composition and transport of salmon calcitonin in transdermal iontophoresis, pp. 63–66, Copyright 1997 with permission from Plenum Publishing Corporation.)

of freeze-dried mixture of salmon calcitonin and gelatin. When required, the disc was placed directly on moistened skin and covered with a wetted pad which allowed immediate dissolution. The delivery system containing the disc reservoir is shown in Figure 5.3. The pad in turn was fixed to the skin by means of an adhesive patch containing the electrode (Santi *et al.*, 1997).

5.5.6 Miscellaneous peptide candidates

In addition to the peptides discussed above, extensive investigations have been done on iontophoretic delivery of insulin. This is discussed in some detail in the following section. In this section, some miscellaneous studies with various peptides are presented. Iontophoretic delivery of octreotide acetate in rabbits has been investigated. Octreotide, a synthetic octapeptide, is a somatostatin analogue commercially available (Sandostatin®, Sandoz) for treatment of acromegaly and carcinoid tumours. Plasma levels of octreotide were negligible without application of current, increased in proportion to current density in the range 50–150 $\mu A/cm^2$ and declined rapidly upon removal of the device. Plasma levels also increased proportionally by increasing drug concentration in the device from 2.5 to 5 mg/ml but not beyond this concentration (Lau *et al.*, 1994). Delivery of octreotide to humans has also been investigated (see Section 8.2). Other polypeptides that have been delivered iontophoretically include an analogue of growth hormone-releasing factor with a molecular weight of 3929. No permeability was observed *in vitro* across hairless guinea pig skin under passive conditions but delivery was achieved by iontophoresis (Kumar *et al.*, 1992). Iontophoretic delivery of parathyroid hormone has been investigated in swine. This polypeptide has applications in osteoporosis. Doses in excess of 25 µg (therapeutic dose) were administered following 4 h of iontophoresis. Peak plasma levels equivalent to those achieved following subcutaneous injection were achieved within 90 min of iontophoresis and fell rapidly when the current was turned off (Boericke *et al.*, 1996). Percutaneously administered elastin peptides were shown to have a strong affinity for rat skin. Resorption through the skin was a slow process, with 30–40 per cent of the administered label still found in the skin after 48 h. Very little or no radioactivity was detectable in the skin at any time in the blood (Menasche *et al.*, 1981).

5.6 Iontophoretic delivery of insulin: fact or fiction?

Insulin, being a protein, cannot be administered orally. It is currently administered by subcutaneous injection but repeated injections are required because of its short half-life. Even with long-acting preparations, the inconvenience of making an injection still exists. Alternative non-invasive delivery systems are therefore desirable and have been reviewed (Chien and Banga, 1989b). This discussion is limited to transdermal delivery. One possibility is iontophoretic delivery. Other techniques such as phonophoresis may also be feasible (Tachibana and Tachibana, 1991; Tachibana, 1992) but are beyond the scope of this discussion. Insulin associated with transferosomes (also see Section 3.5.1) has also been reported to be carried across the skin with an efficacy of 50 per cent or more (Cevc *et al.*, 1995). Research on the transdermal iontophoretic delivery of insulin for systemic effect is relatively recent with the first publication appearing in 1984. Since then, several reports have been published. However, these studies have not yet conclusively established the feasibility of delivering intact insulin across human skin in therapeutically meaningful amounts. This is because different investigators have used different methods and there are several variables to be considered before data can be interpreted. First and foremost, the molecular weight of the polypeptide being administered is not certain. At a molecular weight of about 6000, insulin would be expected to have sufficient permeability under iontophoresis to make the technology feasible. However, most commercially available insulin products actually exist in hexameric form, so that we may really be trying to deliver a protein with a molecular weight of about 36 000, which is most likely too high to be within the scope of iontophoretic delivery. Furthermore, the pI of insulin (5.3) falls in the region of skin pI (4.0 to 6.0). This poses major hurdles to its delivery as discussed in general for peptides (see Section 5.3). While there is no evidence for precipitation of insulin in the skin, the formation of a depot upon iontophoresis of insulin has been suggested by several investigators. However, the depot effect could also result simply from the accumulation of insulin in the less accessible regions of the skin and its subsequent slow leaching from those regions. The self-association behaviour of insulin may also be implicated in its depot effect. While insulin circulates in blood in low concentrations (10^{-8}–10^{-11} M) and brings about its biological effects as a monomer, it dimerizes at higher concentrations and in the presence of zinc ions it further assembles into hexamers (Bi *et al.*, 1984). Thus, insulin exists as a hexamer as commonly used but its absorption may require a monomeric form. Absorption of insulin following subcutaneous injection or infusion has been shown to depend on the volume and concentration of the injected or infused insulin and has been modelled taking into account its aggregation status and binding in tissue (Mosekilde *et al.*, 1989). Slow release of monomeric or dimeric insulin from an accumulation of hexameric insulin in the skin can also account for the depot effect. If monomeric analogues of insulin or formulations in which insulin exists as a monomer are used, it could lead to better absorption from subcutaneous site.

In the first reported study (Stephen *et al.*, 1984) on the transdermal iontophoretic delivery of insulin for systemic effect, it was attempted to deliver regular soluble insulin to human volunteers. Iontophoresis of commercially available insulin was done on eight volunteers but negative results were obtained even after repeating the study on three occasions. However, investigators were able to deliver a highly ionized monomeric form of insulin to one pig and observed a decline in blood glucose levels and an increase in serum insulin levels. This modified insulin, though not discussed by the authors, was presumably sulphated insulin. However, while sulphated insulin (Moloney *et al.*, 1964) has been available since the 1960s, it has not found any widespread use in clinical medicine. In 1986, Kari tried to deliver regular, soluble insulin by iontophoresis to New Zealand white rabbits made diabetic by injection of alloxan. Hair was removed from the back of animals by an electric clipper and the animals were shaved, a process which probably disrupted the stratum corneum, as acknowledged in the discussion. Using currents of < 1.0 mA on a surface area of 6.2 cm^2, regular porcine-zinc insulin was delivered. The blood glucose levels decreased and serum insulin concentrations increased within 1 h of turning the current on, but these changes continued even after the current was turned off, suggesting accumulation of insulin in the skin and subcutaneous tissues (Kari, 1986). The feasibility of iontophoretic delivery of insulin to diabetic rabbits following disruption of the stratum corneum has also been demonstrated in a more recent study (Shin and Lee, 1994). In this study, a bifunctional iontophoretic patch system was used which delivered insulin from both anodic and cathodic chambers. This was made feasible by filling the anodal chamber with positively charged bovine insulin solution, adjusted to pH 3 with acetic acid, and filling the cathodic chamber with negatively charged insulin solution, adjusted to pH 8 with sodium salicylate.

In another study on 26 diabetic albino rabbits, a therapeutic dose of human insulin was reported to be delivered successfully across intact stratum corneum. Active patches contained insulin and an electrical circuit was established when the patch was applied to the skin, delivering a direct current of 0.4 mA over a period of 14 h. Control patches, which had an equal amount of insulin but without the corresponding electrical circuit, were also used. After placement of patches, animals with active patches were found to have a significant elevation in serum insulin levels ($p < 0.05$) with a corresponding reduction in blood glucose levels ($p < 0.01$), while no changes were observed in controls. HPLC analysis of insulin in the patch prior to application, immediately after application, and 9 and 18 h after application indicated that insulin was chemically stable in the patch over the course of the study (Meyer et al., 1989). Several other investigators have demonstrated the *in vivo* iontophoretic delivery of insulin in rats (Siddiqui et al., 1987; Liu et al., 1988; Chien et al., 1990; Sophie and Veronique, 1991; Sun and Xue, 1991). In one study, regular and hairless rats were made diabetic by injection of streptozotocin and purified porcine insulin was iontophoretically delivered using a commercially available Phoresor system. There was some evidence for the formation of an insulin reservoir in the skin, as the blood glucose levels continued to decrease even 2–3 h after the iontophoresis treatment was stopped (Siddiqui et al., 1987). Immunochemical detection of insulin following iontophoretic transport across hairless rat skin suggests that transport pathways are predominantly via a transfollicular route. Thus, after iontophoresis, the sebaceous glands were stained much more strongly than the epidermis (Sophie and Veronique, 1991). Another study using laser scanning confocal microscopy and FITC-insulin suggests that the major permeation barrier is located at the stratum corneum, with iontophoretic enhancement taking place via a paracellular pathway (Chen and Chien, 1997).

It has been suggested that pulsed current is more effective than simple direct current for delivery of insulin, with a higher frequency offering better blood glucose reduction when the same average current density is used (Chien et al., 1987; Liu et al., 1988). However, another study found no difference between the effectiveness of DC or pulsed iontophoresis for delivery of insulin. In this study, a photoetched microdevice with a square anode and a U-shaped inset cathode was used to deliver insulin iontophoretically to normoglycaemic control and diabetic rats. A significant reduction of blood glucose level and an increase in immunoreactive insulin concentration was observed in diabetic rats following cathodic DC iontophoresis (Haga et al., 1997). While most of these studies were done on anaesthetized rats, one study was conducted on conscious, streptozotocin-induced, diabetic fuzzy rats. This study also suggests that DC pulse waveform is more effective that steady DC (Sun and Xue, 1991).

The *in vitro* transport and stability of insulin across ethanol-treated human skin has been investigated. Control experiments without ethanol pretreatment of human skin showed no measurable flux of insulin at applied voltage drops of 0.25 and 0.5 V. The skin was then pretreated with 100% ethanol for 2 h and a voltage of 0.5 V was then applied for 10 h, using a donor containing 7 mg/ml of unlabelled porcine insulin. Gel electrofocusing analysis (isoelectric point determination) of the donor and receptor after iontophoresis and the control resulted in all samples reaching the same position on the gel corresponding to a pH of about 5.2, very close to the isoelectric point of insulin (pI of about 5.3). This suggests that insulin does not undergo any metabolism during transdermal iontophoresis after ethanol pretreatment of skin (Srinivasan et al., 1989). However, ethanol pretreatment could have denatured the proteolytic enzymes. The author has observed extensive degradation of ^{125}I-insulin (porcine) during *in vitro* transport across hairless rat skin (Banga and Chien, 1993c), but again, proteolytic activity of animal skin is typically higher than that of human skin.

In another study which used freshly excised rabbit inner pinna skin, the addition of skin to a control buffer increased insulin degradation from 9 to 40 per cent during the first 8 h. When current was applied, the degradation was 70 per cent during the first 8 h. All insulin was completely degraded within 24 h irrespective of the conditions used. In addition, almost all insulin was degraded as it passed through the skin during iontophoresis (Huang and Wu, 1996b). Insulin has been reported to undergo degradation at the injection site in insulin-resistant diabetics and skin is known to contain both exo- and endopeptidases, with the aminopeptidases being the best known (Hopsu-Havu *et al.*, 1977; Banga and Chien, 1993b). Insulin has also been shown to denature under an electric field, with the extent of denaturation depending on the intensity of and duration of exposure to the electric field. The electrochemical stability of insulin analogues was found to depend on the type of amino acids and location of substitution (Chen *et al.*, 1997a, b).

The observation that no permeation was seen across untreated human skin in the study discussed earlier (Srinivasan *et al.*, 1989) under the conditions used apparently conflicts with the several published reports of successful iontophoretic delivery of insulin across animal skin. This may be due to the fact that human skin is known to be less permeable than animal skin. As discussed, monomeric analogues of insulin may be better candidates for its iontophoretic delivery. It has been shown that clinically relevant amounts of insulin could be iontophoretically delivered, after wiping the skin with absolute alcohol. In this study, a series of monomeric analogues of human insulin could be iontophoretically delivered across hairless mouse skin, while only low and insignificant flux was observed with regular hexameric human insulin (Langkjaer *et al.*, 1994). A more important modification may be the development of insulin analogues with modified pI. In a study with several insulin analogues, all analogues with a pI inside the pH range of skin resulted in undetectable flux. However, sulphated insulin (pI ~ 2), an analogue with a pI outside the skin pH range, did show measurable flux in a porcine skin flap model. When cathodal iontophoresis was conducted for 4 h at 0.4 mA using a 2-cm^2 reservoir, the total recovery was about one-quarter of a unit of insulin. Under these conditions, the biological response in streptozotocin-induced diabetic rabbits resulted in a significant glucose decline of 200–300 mg/dl, proving that at least some biological activity was retained in the sulphated insulin delivered iontophoretically. Calibration against a subcutaneous dose and comparison with skin flap data suggested that a large fraction of the dose was biologically active (Sage *et al.*, 1995). Another study has shown a voltage-dependent enhancement in the iontophoretic permeation of insulin across human cadaver skin. Enhancers such as DMSO, urea and bile salts were also able to increase the flux (Rao and Misra, 1994). Reverse iontophoresis (see Section 2.7) can be exploited for non-invasive glucose measurement by its iontophoretic extraction from the subcutaneous tissue (Rao *et al.*, 1993). While this step itself has tremendous commercial value for non-invasive glucose monitoring, the next logical step would be to link this glucose measurement to a closed loop biofeedback system for iontophoretic delivery of insulin.

As is evident from the above discussion, extensive work needs to be performed to make transdermal iontophoretic delivery of insulin a viable technique for the treatment of diabetics. The *in vivo* studies are complicated by biological variations as blood glucose and plasma insulin levels can fluctuate due to several factors including, but not limited to, circadian rhythms, anaesthesia and stress. Also, these studies were conducted in laboratory animals and results may not be extrapolated to human skin. For example, rabbit skin could contain a very high hair follicle density, a crucial difference since iontophoretic transport takes place through these shunt pathways (Banga and Chien, 1988). On the other hand, poor experimental design and the lack of sensitive analytical techniques used for most *in vitro* studies failed to assess the stability of insulin during transport. A significant portion of radioactivity could very well be insulin degradation products instead of the intact, labelled insulin. The depot effect observed by several investigators could be a disadvantage for modulation of delivery or development of a biofeedback system. However, iontophoresis would still be useful to load skin tissues with insulin in a non-invasive manner, for a relatively prolonged release similar to the long-acting preparations on the market. The use of other electrical assistance mechanisms, such as electroporation, has not been investigated to enhance the transdermal delivery of insulin. The combined use of iontophoresis and electroporation may also result in some interesting results. Other physical mechanisms to assist permeation, such as the use of ultrasound, may also be promising. The use of penetration enhancers or other novel formulations such as liposomes can also be investigated. A recent publication has suggested that porcine insulin, following dissociation to a dimer in a 0.1 M glycine-HCl buffer (pH 4.0), was absorbed *in vivo* through Wistar rat skin, as estimated by blood glucose levels. The best absorption was obtained with a liposomal formulation containing two penetration enhancers (Ogiso *et al.*, 1996). The combination of iontophoresis with penetration enhancers may also be feasible, though it is likely to result in a complex system that is difficult to characterize. In a recent study, 5% azone/propylene glycol increased the iontophoretic steady-state flux of insulin by a factor of 2.75 compared with iontophoresis alone. In this study, full-thickness skin from the back region of a mouse was used after the hair was removed with an electric clipper (Hao *et al.*, 1996).

6

Transdermal delivery by electroporation

6.1 Introduction

An electric enhancement technique that could work alone or in conjunction with iontophoresis is electroporation, which involves the application of a high voltage pulse for a very short duration. Electroporation is best known as a physical transfection method in which cells are exposed to a brief electrical pulse, thereby opening pores in the cell membrane, allowing DNA or other macromolecules to enter the cell (Treco and Selden, 1995; Weaver, 1995). Electroporation reversibly permeabilizes lipid bilayers, and possibly involves the creation of aqueous pores during the application of an electric pulse. The formation of aqueous pathways ('pores') which are straight-through with a radius of a few nanometres is more energetically favourable than long, tortuous pathways around corneocytes (see Figure 1.1). During electroporation, the transmembrane potential produces a current which passes through the membrane defects and leads to a reversible increase in permeability. The exact mechanism for electroporation is not clear though the pore mechanism is generally believed to be the case (Tomov, 1995). Changes in the behaviour of membranes seen following electroporation are consistent with the theory of pore formation. These include changes in electrical or mechanical behaviour or those seen in molecular transport. Changes in membrane conductance are often dramatic and are believed to be due to ionic conduction through transient aqueous pores. However, these pores have not been visualized by any microscopic techniques, presumably owing to factors such as their small size and transient nature. Electroporation is considered to be a non-thermal phenomenon as pore formation by membrane rearrangement occurs much before any significant temperature rise takes place in the pulsing medium (Weaver, 1993). Electroporation is also related to another event called electrofusion. If two cells are brought into close membrane–membrane contact and then electroporated, the membranes of these two cells may collapse at their point of contact, resulting in cell fusion and cytoplasmic mixing (Radomska and Eckhardt, 1995; Jones *et al.*, 1996). The main use of electroporation is for the introduction of DNA into cells. This topic is outside the scope of this book but is discussed briefly in Chapter 7. Electroporation as a science has about 25 years of history

(Tsong, 1991) though its use for transdermal or topical delivery was suggested only about 5 years ago (Prausnitz et al., 1993a). This chapter will discuss the use of electroporation for electrically assisted transdermal or topical delivery of drugs other than oligonucleotides or DNA-based drugs. Applications of electroporation for delivery of oligonucleotides, genes and DNA-vaccines are discussed in Chapter 7.

The technique of electroporation is normally used on the unilamellar phospholipid bilayers of cell membranes. However, it has been demonstrated that electroporation of skin is feasible, even though the stratum corneum contains multilamellar, intercellular lipid bilayers with few phospholipids (Prausnitz et al., 1993a). The electrical behaviour of human epidermal membrane as a function of the magnitude and duration of applied voltage closely parallels the electrical breakdown/recovery of bilayer membranes seen during electroporation (Inada et al., 1994). The approximately 100 multilamellar bilayers of the stratum corneum need about a 100-V pulse for electroporation, or about 1 V per bilayer (Weaver and Chizmadzhev, 1997). The stratum corneum does not contain any living cells, but it can still be permeabilized by an electric pulse because electroporation is a physical process based on electrostatic interactions and thermal fluctuations within fluid membranes and no active transport processes are involved. A series of experiments using two molecules of similar size but different charge (calcein, 623 Da, charge = -4; and sulphorhodamine, 607 Da, charge = -1) showed that their electrical behaviour and transport were consistent with the hypothesis that aqueous pathways are caused in skin by high voltage pulsing (Pliquett and Weaver, 1996a). Transport of calcein through human stratum corneum during electroporation was shown to occur through intercellular and transcellular pathways (Prausnitz et al., 1996a). While iontophoresis involves the use of relatively low transdermal voltages ($\ll 100$ V), electroporation of skin takes place at high transdermal voltages (~ 100 V or more). There is considerable indirect evidence that high voltage pulses cause changes in the skin structure (Edwards et al., 1995; Prausnitz, 1996b).

The use of electroporation in conjunction with iontophoresis can expand the scope of transdermal delivery to larger molecules such as therapeutic proteins and oligonucleotides. While iontophoresis acts primarily on the drug, electroporation acts on the skin with some driving force on the drug during a pulse (Prausnitz et al., 1993a). However, iontophoresis will have secondary effects on the skin just like electroporation would also apply direct electromotive force on the drug during the brief pulse period. This may be particularly true for the thin cell lining of the sweat ducts, which might be electroporated with low voltages of the order used in iontophoresis. Electroporation has been shown to increase the flux of LHRH significantly and reversibly through human skin. The application of a single pulse prior to iontophoresis increased the flux by 5–10 times over that achieved by iontophoresis alone (Bommannan et al., 1994). While iontophoresis can be used to provide baseline levels, electroporation pulses can potentially be applied to provide rapid boluses. This has been shown to be feasible for transdermal transport of calcein across human epidermis. More complex delivery schedules were also achieved by changing pulse voltage (Prausnitz et al., 1994). Thus, the combined use of iontophoresis and electroporation is likely to yield useful and interesting data which will intensify the efforts to explore electroporation more fully as a means of transdermal drug delivery. One advantage of giving an electroporation pulse in conjunction with iontophoresis for proteins may be that the aggregation status of the protein may become less important for its delivery across or into skin as electroporation can potentially deliver drugs of higher molecular weight than iontophoresis.

6.2 Electropermeabilization of the skin

The electrical properties of the skin have been discussed in Section 2.2. Above a certain threshold value, the resistance of the skin is progressively reduced. A retrospective investigation using data from several studies has shown that the resistance of the skin depends most strongly on voltage, followed by current density and power density. Since electroporation involves the use of very short pulses, much higher voltages (compared with iontophoresis) can be used without causing sensation. The minimum current that is required to be applied to the skin to evoke a sensation is termed the perception threshold, while the minimum to cause a painful sensation is called the pain threshold. The latter is more variable as it depends on both physical and sociological factors and is typically 3–25 times greater than the former. The perception and pain thresholds do not scale directly with either current (I) or current density (Ij) but exhibit more complex functionalities. Power-function least-squares fit of the data resulted in the following relationships for perception thresholds:

$$I = 0.00042\ A^{0.21}\ (r^2 = 0.47)$$

$$Ij = 0.00042\ A^{-0.79}\ (r^2 = 0.93)$$

and the following for pain thresholds (Prausnitz, 1996a, 1997):

$$I = 0.0033\ A^{0.33}\ (r^2 = 0.45)$$

$$Ij = 0.0033\ A^{-0.67}\ (r^2 = 0.77)$$

where A is the electrical contact area. For single pulses of about 90 V across human stratum corneum, the recovery of the skin resistance has been shown to be very complete, returning to about 90 per cent of the pre-pulse value. The recovery consisted of four phases, with about 60 per cent recovery in phase I (20 ms), 40–70 per cent in phase II (0.4–0.8 s) and 60–90 per cent in phase III (10 s). This is followed by a slow phase of recovery of over a minute to a few hours. For higher voltage pulses (> 130 V), the recovery was typically less than 50 per cent (Pliquett *et al.*, 1995a). The recovery time for these conditions may be quicker if full-thickness skin is used. Using excised full-thickness porcine skin, the recovery was shown to take place almost instantaneously or within the time required to switch to the measuring instrument. This was claimed to be due to the stratum corneum being still attached to the epidermis and underlying tissue, and being relaxed, unlike the former study where the stratum corneum was heat stripped, hydrated and mounted in a chamber. The permeabilization was found to depend on the electrical exposure dose, which is the product of the pulse voltage and the cumulative pulsing exposure time. Skin resistance was observed to drop to 20 per cent of its pre-pulsing value when pulsed beyond a critical dosage of 0.4 V/s, but recovered rapidly. When the dose exceeded 200 V/s, the recovery was slow and incomplete (Gallo *et al.*, 1997).

As mentioned earlier, electroporation is believed to act on the skin, with direct electrophoretic effects on the drug contributing relatively less. Support for this comes from several studies. For instance, it has been shown that application of continuous low voltage resulted in a calcein flux three orders of magnitude lower than pulsing at high voltage under 'electrophoretically equivalent' conditions, suggesting that structural changes induced in the skin by pulsing contribute more significantly than the direct electrophoretic force acting on the drug (Prausnitz *et al.*, 1993a). Calcein, being a fluorescent molecule, has also allowed measurement of rapid kinetics of transdermal transport. The measured fluorescence is a function of the rate of transdermal transport, but as continuous measurements are being made, the reading may also be a function of mixing and pumping times of the measurement systems. After deconvoluting the signal, it was suggested that the transport due to each pulse occurred over a time scale in the order of 10 s or less (Pliquett *et al.*, 1995b). Transport of calcein across erythrocyte ghost membranes using electroporation has also been measured with millisecond time resolution. It was observed that while much of the transport is electrically driven, transport can also occur by diffusion within seconds after a pulse (Prausnitz *et al.*, 1995a). The calcein flux reached steady-state levels within minutes and then decreased below the detection limit within seconds following the termination of a pulse. While flux was dependent on both pulse voltage and pulse rate, the steady-state lag time and onset time were found to depend only on the pulse rate (Prausnitz *et al.*, 1994). In contrast, the flux of a macromolecule, heparin, remained partially elevated for several hours after high voltage pulsing (Prausnitz *et al.*, 1995b).

A chemical enhancer effect for transdermal transport has been shown by high voltage pulsing in the presence of macromolecules. Heparin has been shown to alter its own transport during electroporation (Prausnitz *et al.*, 1995b) as well as the transport of other molecules (Vanbever *et al.*, 1997; Weaver *et al.*, 1997). It has been hypothesized that heparin or other linear macromolecules (such as DNA) can enter an aqueous pathway between corneocytes, become trapped, and keep the pathways open. This resulted in a persistent, low, post-pulse electrical resistance of the skin in the presence of heparin (~ 300 Ω). In the absence of heparin, the post-pulse recovery was from the initial post-pulse 400 Ω to about 800 Ω within 2 h (pre-pulse skin resistance was about 82 kΩ). As a result of these open pathways, transport of sulphorhodamine (charge = -1) was increased in the presence of heparin. However, the transport of highly charged (charge = -4) calcein was decreased, presumably because of the electrostatic repulsion between the highly negatively charged calcein and the negatively charged heparin trapped in the aqueous pathways (Weaver *et al.*, 1997). In a recent study, several macromolecules such as dextran sulphate, neutral dextran and poly-lysine were shown to have an enhancer effect on the electroporation-induced transdermal delivery of mannitol across freshly excised abdominal hairless rat skin. Macromolecules with a greater charge and size were found to be more effective enhancers. The effect of macromolecules added at the time of pulsing lasted for several hours after pulsing. It has been suggested that these enhancers do not disrupt lipids but instead stabilize the transient disruptions already induced by electroporation of skin as a result of their flexible linear structures entering the skin. This hypothesis was supported by the observation that an enhancer effect was seen only for electroporation, and not for passive diffusion or iontophoresis. Small ions were also used but provided no enhancement for electroporation (Vanbever *et al.*, 1997).

6.3 Electroporation equipment

Electroporation equipment is typically manufactured for genetic manipulation of living cells, and has been used as such or adapted for transdermal studies. Commercially available equipment will come with a chamber which is usually a sealable cylinder with an electrode plate at each end. These chambers are typically disposable, presterilized cuvets with moulded-in aluminum electrodes and come in different gap sizes, depending on whether bacteria or mammalian cells need to be electroporated (Hofmann, 1995a; Jones et al., 1996). The chambers are not required for electroporation of tissues and will not be discussed here. In such studies, the nominal field strength, E, is related to the applied voltage, V, and the interelectrode gap, d ($E = V/d$). Field strengths vary from a few hundred volts per centimetre for mammalian cells to many kilovolts per centimetre for bacteria. Typically, electroporation equipment is bought as a commercially available model, though it is possible to build one if the person has sufficient electronics expertise to ensure safety of the user under all operating conditions (Jones et al., 1996). A capacitor is connected to a power supply to build up an electric charge and is then discharged after isolating from its charging source. The current and pulse length obtained depend on the speed at which the stored energy is released, which in turn can be controlled by using resistors (Jones et al., 1996). The voltage output from electroporation equipment may be in the form of exponential decay or square-wave pulses. The exponential decay waveform represents complete discharge of the capacitor into a resistor, while a square-wave pulse is often produced by a partial discharge of a large capacitor. For an exponential waveform generator, the voltage of a capacitor C discharging into a resistor R follows an exponential decay law:

$$V = V_0.\exp(-t/RC)$$

The pulse length will be characterized by the 1/e time constant, which is the time required for the initial voltage to decay to 1/e, approximately one-third of the initial value. This time constant is a product of R and C, where C is the capacitance of the storage capacitor in the generator and R is the total resistance into which the capacitor discharges (Hofmann, 1995a). Both waveforms have been used successfully for electroporation of skin, though it has been suggested that exponential decay pulses may be somewhat more effective than square-wave pulses at the same applied energy (Vanbever et al., 1996). However, enough studies have not been done to establish unequivocally the superiority of one pulse waveform over the other. The waveforms are determined by the principles of electrical engineering used by the pulser, which generally have to be designed for one waveform only. Unlike capacitive discharge devices, a microprocessor-controlled logic-driven unit can produce reproducible DC pulse lengths independent of the conductivity of the solutions bathing the cells (Jones et al., 1996).

Based on the information provided in the literature (Hofmann, 1995a) and available on the Internet, several commercial brands of electroporation equipment can be identified. A partial list would include pulsers made by the BTX division of Genetronics, Inc. (San Diego, CA, USA), Cyto Pulse Sciences (Columbia, MD, USA), as well as several units from IBI (New Haven, CT, USA), Invitrogen (San Diego, CA, USA), Bio-Rad (Richmond, CA, USA), BRL (Grand Island, NY, USA), and Equibio (Kent, UK). The Electro cell manipulator (ECM) 600 (Figure 6.1) from Genetronics is an exponential decay electroporator that can switch between a high and a low voltage mode with the turn of a knob. The high voltage mode is for bacterial transformation, but the low voltage mode is designed for plant or mammalian transfection and can be used for skin electroporation. The instrument has 7 capacitors and 10 resistors allowing a wide choice of settings for the pulse length. Pulse Agile™ from Cyto Pulse Sciences is capable of setting the parameters of each individual pulse in any square-wave pulse sequence.

Figure 6.1 The Electro cell manipulator (ECM) 600. (By courtesy of Genetronics, Inc., San Diego, CA, USA.)

6.4 Factors affecting delivery by electroporation

Key variables in the enhancement of percutaneous absorption of drugs by electroporation are the selection of the pulse voltage, pulse duration and the number of and spacing between pulses. The goal is to maximize drug delivery using a protocol which will be acceptable for *in vivo* human studies. For *in vitro* experiments, significant voltage drop occurs within donor and receptor solutions, so that applied voltages need to be significantly higher (Prausnitz *et al.*, 1994). The application of high voltage electric pulses may be accompanied by temperature changes in the pulsing medium due to electrical power dissipation or even interfacial electrochemical heat production (Pliquett *et al.*, 1996). Thus, results obtained under different conditions should be interpreted cautiously. The use of short pulses adding up to the duration of a longer single pulse will help to minimize the heating. Since the bulk of research with electroporation has been done on cells and not tissues, literature relating to the former can be important in shedding some light on the factors involved. A quantitative dependence on the percentage of porated cells and the parameters of the pulse such as shape, amplitude and duration has been derived and the expressions compared well with experimental results on the electroporation of human erythrocytes (Tomov, 1995). The voltage used to electroporate cells is typically very high. However, recently, low voltage electric fields (about 60–80 V cm^{-1}) have been reported to incorporate macromolecules into cells and vesicles. The mechanism was not electroporation as the electric field was too low to induce permeability changes of the membrane. Instead, it was hypothesized that electrophoresis induced an increase in charged macromolecules on the surface of the cells which interacted with the plasma membrane of the cells to destabilize the membrane and the resulting membrane vesiculation led to an endocytotic-like uptake (Rosemberg and Korenstein, 1997). For studies with skin, a typical applied voltage used is 100–300 V (Vanbever et al., 1996), though voltages as high as 1000 V have been used during in vitro transdermal studies. The electric field represents the voltage (V) that is applied across the electrode gap (cm). For instance, if a pair of electrodes is put across a 5-mm tumour and a voltage of 500 V is applied, then the strength of the electric field is 500 V/0.5 cm = 1000 V/cm or 1 kVcm (Dev and Hofmann, 1994).

For the transport of calcein (623 Da) across human epidermis, it was found that flux increased as a strong function of voltage up to 100 V, but only weakly above this value. Transdermal flux of calcein was enhanced by up to four orders of magnitude by pulsing for 1 h. After the pulsing was stopped, the flux decreased by about 90 per cent within 30 min and by more than 99 per cent within 1 or 2 h, indicating reversibility when the voltage was at or below about 100 V (Prausnitz et al., 1993a). It has been suggested that a large number of high voltage–short duration pulses may permeabilize the skin efficiently, but may be less efficient than a small number of low voltage–long duration pulses to obtain an electrophoretic movement of drug (Vanbever et al., 1996). However, a long duration pulse may be associated with increased skin irritation. In a study on the transport of metoprolol, a linear correlation was observed between pulse voltage (24–450 V; pulse time 620 ms) and cumulative metoprolol transported after 4 h. A linear correlation was also observed with pulse time (80–710 ms) at 100 V (Vanbever et al., 1994). For studies with calcein, the transport increased almost linearly with transdermal voltage above a threshold of about 80 V and then levelled off at higher voltages (> 250 V) or shorter spacing (< 10 s) between pulses (Pliquett and Weaver, 1996b). If desired, the voltage traces such as pulsing voltage or transdermal voltage may be acquired and stored on a digital oscilloscope (Zewert et al., 1995; Heller et al., 1996; Prausnitz et al., 1996b). The current flowing through the skin during the first pulse will be different from subsequent pulses because the resistance of the skin will drop. Also, when using in vitro diffusion cells, there will be a significant drop in voltage within the donor and receptor solutions.

The electrodes used for pulsing in transdermal studies have mostly been silver/silver chloride. The use of silver/silver chloride electrodes in iontophoresis has been discussed in Section 3.4.2. The material should be such that it can accommodate a high instantaneous charge density and should not form harmful electrochemical products. Homogeneously mixed silver/silver chloride electrodes may be best since silver wires electrochemically plated with silver chloride may be susceptible to the detachment of the outer layer of coating during pulsing. While wire electrodes have been used for in vitro studies, more elaborate designs are required for in vivo studies. One possible design for in vivo tissue electroporation has been described in the literature (Prausnitz et al., 1993b). A combination of a stainless steel cathode and a silver anode has also been used to provide an inexpensive alternative applicable to the particular situation. A second pair of inner electrodes located closer to the skin were made of silver/silver chloride and were used for electrical measurements because of their low electrode polarization and phase stability (Pliquett and Weaver, 1996a). The design of electrodes for use in cancer chemotherapy is discussed in Section 6.8. The effect of competitive ions and electrode polarity may be less significant in electroporation if passive permeation of the drug through electropermeabilized skin represents a significant percentage of flux. Delivery of fentanyl following electroporation of hairless rats was not affected significantly by a 10-fold change in the buffer concentration or by reversal of electrode polarity (Vanbever et al., 1996). However, in a meander electrode set-up (see Section 6.7), increase in ionic strength may result in increase in shunting of current across the finger electrodes. The delivery of a drug by electroporation may be hindered by the presence of high concentrations (> 15% v/v) of cosolvents (such as propylene glycol) in the formulation. This could be due to a decrease in the conductivity of the drug solution (Jadoul et al., 1997).

6.5 Pulsatile delivery of drugs by electroporation

Besides the model compounds calcein, sulphorhodamine and caffeine which are discussed elsewhere in this chapter, other drugs which have been investigated for transdermal delivery by electroporation include fentanyl (Vanbever et al., 1996) and metoprolol (Vanbever et al., 1994). Results with many of these drugs have been less dramatic than those seen with model compounds. While studies with model compounds have given excellent mechanistic insights, the magnitude of flux enhancement observed may not happen for all drugs so that each drug needs to be studied as a separate case. For the delivery of the nonsteroidal anti-inflammatory drug flurbiprofen, both iontophoresis and electroporation were effective in rapid delivery of the drug into the systemic circulation of male Wistar rats. The same total charge (1.3 C) and duration of treatment was used for both techniques. The average (± SEM) area under the flurbiprofen plasma concentration–time curve (AUC) was 87.1 ± 20.5 µg.h/ml by iontophoresis, 110.5 ± 18.0 µg.h/ml by electroporation, and 91.7 ± 4.3 µg.h/ml following intravenous bolus (Cruz et al., 1997). Delivery of genes by electroporation is discussed separately in Chapter 7. Delivery of cyclosporin A by electroporation has also been reported. The drug is useful for treatment of psoriasis but is toxic if given systemically and no topical formulations currently exist. By using a single pulse at a field strength of 200 V/cm and a 10 ms pulse interval, the delivery through rat skin was enhanced by a factor of 8.5 over passive diffusion (Wang et al., 1997). Some studies on LHRH and heparin are discussed in the following paragraphs.

6.5.1 LHRH

As discussed, the iontophoretic delivery of luteinizing hormone-releasing hormone (LHRH) across epidermis separated from human cadaver skin has been shown to be significantly enhanced following a single electroporation pulse. The pulse was an exponentially decaying type with an initial amplitude of 1000 V and a time constant of 5 ± 1 ms. At a current of 0.5 mA/cm^2, the flux was 0.27 ± 0.08 without the pulse and increased to 1.62 ± 0.05 µg.h/cm^2 with the pulse (Bommannan et al., 1994). The usefulness of electroporation in enhancing the iontophoretic flux of LHRH has been verified using the isolated perfused porcine skin flap (IPPSF) model, a model which closely resembles human clinical use (see Section 3.3.3). It was found that the application of a single pulse (500 V, 5 ms) immediately prior to 30 min of iontophoresis increased the LHRH concentration in the IPPSF perfusate by nearly twofold, while application of a pulse every 10 min resulted in a threefold increase (Riviere et al., 1995). By using repeated applications of the pulse/iontophoresis protocol, it was shown that electroporation is able to enhance LHRH transport repeatedly in a pulsatile manner relative to iontophoresis (Figure 6.2), thus allowing pulsatile delivery of therapeutic peptides.

6.5.2 Heparin

Heparin is a class of molecules (MW 5000–30 000) where smaller molecules have a weaker anticoagulation activity. It is used clinically for anticoagulation and prophylaxis of

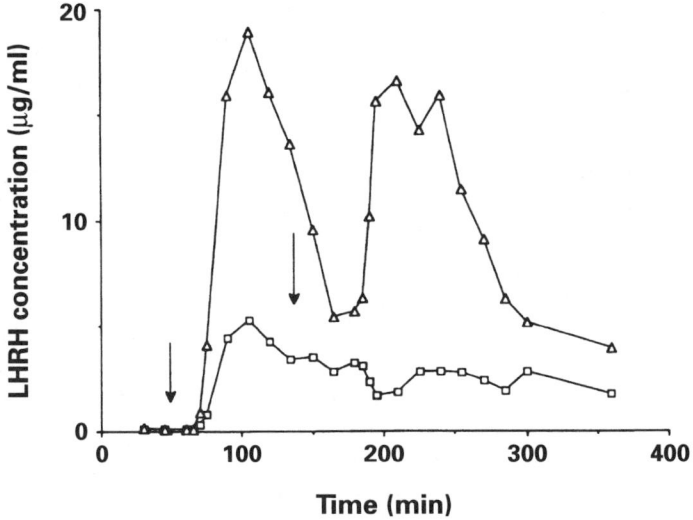

Figure 6.2 A representative pair of IPPSFs in which LHRH was delivered in two separate episodes consisting of either one 500 V pulse followed by 30 min of iontophoresis (△), or 30 min of iontophoresis alone (□). The arrows indicate the beginning of each treatment period. (Reprinted from *Journal of Controlled Release*, **36** Riviere *et al.*, Pulsatile transdermal delivery of LHRH using electroporation: Drug delivery and skin toxicology, pp. 229–233, Copyright 1995 with kind permission from Elsevier Science – NL, Sara Burgerhartstraat 25, 1055 KV Amsterdam, The Netherlands.)

thromboembolism. It is not available upon oral administration and continuous infusion is required for full-dose therapy. Thus, transdermal delivery would be useful. It can also have application for topical therapy of vascular permeability diseases, superficial thromboses, inflammatory and arthritic conditions. Owing to heparin's high molecular weight, skin permeation needs to be enhanced by a suitable mechanism such as penetration enhancers (Xiong *et al.*, 1996) or electroporation. Transdermal transport across human skin *in vitro* has been shown to be feasible at therapeutic rates (100–500 μg.h/cm^2) by applying short (pulse length 1.9 ms), high voltage (150–350 V) pulses at the rate of 12 pulses/min for 1 h. Heparin retained its biological activity as it was transported across the skin, but as smaller molecules were preferentially transported, the anticoagulant activity of molecules transported was lower than those in the donor compartment. The fluxes achieved by low voltage iontophoresis having the same time-averaged current were an order of magnitude lower than those achieved with electroporation (Prausnitz *et al.*, 1995b). The effect of heparin on transport pathways has already been discussed in Section 6.2.

6.6 Skin toxicology of electroporation

Red blood cells that have been electroporated to insert molecules have long circulating half-lives. The lack of rapid clearance suggests that a significant long-lived change in their membrane has not taken place (Weaver, 1993). Clinical precedents for safely applying electrical pulses of hundreds of volts to the skin exist with the use of techniques such as 'transcutaneous electrical nerve stimulation' (TENS) (Prausnitz *et al.*, 1993a). TENS is commonly used to treat chronic conditions such as low back pain, arthritis and pain caused by a variety of neurological disorders (Prausnitz, 1996a). Its use for acute conditions such as labour pains (Carroll *et al.*, 1997), delayed-onset muscle soreness in humans (Craig *et al.*, 1996), in neck muscles (Pizzamiglio *et al.*, 1996) and other uses (Stevens *et al.*, 1996; Torry *et al.*, 1996) is under investigation. The range of sensations evoked by TENS has varied from tactile (touch, vibration, etc.) to pricking pain and itching. Very small changes in stimulus parameters are needed to convert tactile sensations to pain and vice versa, though thermal sensations are rarely reported (Garnsworthy *et al.*, 1988). Nevertheless, electroporation of skin or tissue is a relatively new area and much still needs to be learned about its safety. Electroporation of the back skin of anaesthetized hairless rats has been reported to result in direct stimulation of motor nerves. As each pulse was given, the hind legs of the rat were observed to kick and the intensity of kicking varied with the applied voltage and electrode position (Prausnitz *et al.*, 1993b). For electrochemotherapy of terminally ill patients (see Section 6.8), electric fields as high as 1.3 kV/cm have been used after 1% lidocaine was injected around each nodule. Muscle contractions were observed and mild pain was felt during each pulse but then subsided at the end of each pulse (Heller *et al.*, 1996). The toxicology of exposure to currents or voltages much higher than those used in electroporation may also give some insights into understanding mechanistic toxicological details. Common electrical injuries have been proposed to involve extensive electroporation of muscle cells. In cardiac procedures comprising electric shock, electroporation of muscle cells and the accompanying Ca^{2+} leakage is a source of clinical complications (Tsong, 1991). Poloxamer 188 has been found to adsorb to lipid bilayers and modify the surface properties. It thus decreases their susceptibility to electroporation and may contribute towards decreasing damage sustained by cells if they are inadvertently exposed to an electric shock (Sharma *et al.*, 1996). Repeated electrical stress will have an adverse effect on epidermal mitosis (Gauthier, 1996).

As the stratum corneum has a much higher electrical resistance than other parts of the skin, an electric field applied to the skin will concentrate in the non-viable stratum corneum to induce electroporation. In contrast, the field will be much lower in the viable tissues, thereby protecting the already permeable viable parts of the skin (part of the epidermis and all of the dermis) and deeper tissues. The reversibility of permeation following electroporation suggests that the technique is not damaging to the skin. As discussed, diagnostic and therapeutic applications such as TENS safely apply electric pulses to skin with voltages up to hundreds of volts and duration up to milliseconds. Long-term toxicological implications of the use of this technique, supported by histological examinations, are currently under way. As the promise of this technique is demonstrated, such studies will be conducted by several groups active in this area before FDA approval for commercialization is sought. The skin toxicology of electroporation relative to iontophoresis has been studied using 14 pigs. In the first study using eight pigs, exponential voltage pulses were applied followed by constant current anodal iontophoresis. The erythema, oedema and petechiae observed under the conditions of study were recorded. Pulses of 0, 250, 500 and 1000 V were applied followed by iontophoresis of 0, 0.2, and 2.0 mA/cm^2 for 30 min or 10 mA/cm^2 for 10 min. The results of gross evaluation immediately after or 4 h after treatment are shown in Table 6.1. As seen from the data, the erythema increases immediately after treatment with increasing pulse voltage, but was absent or minimal after 4 h. It was reported that it disappeared or reduced within 5 min. The pulse voltage had no effect on oedema or petechiae. Erythema, oedema and petechiae all increased with increasing current, though the application of a pulse did not increase the irritation induced by iontophoresis. These changes were comparable with those seen with iontophoresis alone. In the second study with six pigs, it was found that at both the gross and light microscopic

Table 6.1 Gross observations of the skin under the active electrode

Pulse (V)	Erythema				Oedema				Petechiae			
	0.0	0.2	2.0	10.0	0.0	0.2	2.0	10.0	0.0	0.2	2.0	10.0
	(Current, mA/cm^2)											
0	0.00	1.75	1.25	1.75	0.00	1.75	1.25	1.75	0.00	0.75	0.75	1.50
	0.00	0.50	0.50	0.50	0.00	1.00	1.00	1.50	0.00	0.00	0.00	0.50
250	0.00	0.75	1.25	2.00	0.00	0.75	1.50	1.67	0.00	0.00	0.50	1.67
	N/A	0.00	0.50	0.00	N/A	0.50	1.00	1.00	N/A	0.00	0.50	1.00
500	1.00	1.25	1.75	2.30	0.00	1.00	1.50	1.33	0.00	0.00	1.50	1.33
	N/A	0.00	0.50	0.00	N/A	1.50	1.00	1.00	N/A	0.00	0.00	0.00
1000	2.60	1.25	1.50	2.00	0.80	1.25	1.00	1.75	0.00	0.00	0.25	1.50
	1.67	1.00	0.00	1.00	0.67	0.50	1.00	1.00	0.00	0.00	0.50	0.50

The first set of data at each pulse voltage is immediately after treatment, and second set is after 4 h. (Reprinted from *Journal of Controlled Release*, **36** Riviere *et al*., Pulsatile transdermal delivery of LHRH using electroporation: Drug delivery and skin toxicology, pp. 229–233, Copyright 1995 with kind permission from Elsevier Science – NL, Sara Burgerhartstraat 25, 1055 KV Amsterdam, The Netherlands.)

level, electroporation does not result in any skin changes not previously seen with iontophoresis alone (Riviere *et al*., 1995).

6.7 Delivery of particulates into the skin

'Electroincorporation' is a technique in which a drug encapsulated in vesicles or particles is delivered into the skin (see Figure 1.1) by applying a pulse which causes a breakdown of the stratum corneum (Hofmann, 1995b). The technique involves placing particles on the skin and then pulsing with electrodes placed directly on top of the particles. This creates an electric field which breaks down the stratum corneum by a yet unknown mechanism. Slight pressure is applied on the electrode during pulsing. Following breakdown of the stratum corneum, dielectrophoresis and/or pressure is believed to drive the particles into the skin. Pulsing can be accomplished by using a meander electrode (Figure 6.3), which consists of an array of interweaving electrode fingers. This type of electrode configuration is suitable for including in a patch or device. Electrodes shaped like callipers are also available and have been used to pulse a back skin fold of hairless mouse. The particles do not have to be charged for delivery. Particles of 0.2, 4.0 and 45.0 µm in size were shown to be imbedded in hairless mouse skin when pulsed with three exponential decay pulses of amplitude 120 V and pulse length 1.2 ms (Hofmann *et al*., 1995). Pressure-mediated electroincorporation has also been used to deliver Lupron Depot® (leuprolide acetate) microspheres into hairless mouse skin and into human skin xenografted on immunodeficient nude mice. For the deposition of microspheres into the skin, the coverage of the particles on the electrode needs to be considered. For hairless mice, 1.1 million particles were placed on the skin to provide 100 per cent coverage and the resulting efficiency of delivery was 36 per cent. For human skin grafts, 4 million particles were used to provide 300 per cent coverage which reduced the delivery efficiency to 4 per cent. If the number of particles on the skin graft is reduced to provide only

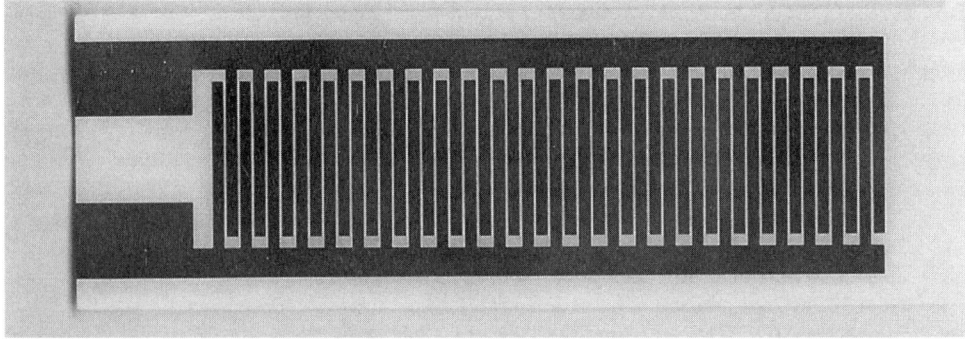

Figure 6.3 Meander electrode which consists of an array of interweaving electrode fingers. (By courtesy of Genetronics, Inc., San Diego, CA, USA.)

monolayer coverage, then delivery efficiency could possibly be higher (Zhang *et al.*, 1997). Similarly, gold particles have been used to enhance the transdermal delivery of caffeine through full-thickness human cadaver skin via pressure-mediated electroincoporation. In this study, 12 exponential decay pulses (120 V, 8 ms) were administered using the meander electrodes (Zhang and Hofmann, 1997). The use of calliper electrodes for delivery of genes is discussed in Chapter 7.

6.8 Electroporation for cancer chemotherapy

A novel form of cancer treatment, called electrochemotherapy, utilizes a combination of electroporation and chemotherapeutic agents (Dev, 1994; Dev and Hofmann, 1994). The technique typically involves the systemic administration of an anticancer drug followed by delivery of electric field pulses at the site of the cancer. The rationale for this approach is that many cancer drugs are very poorly permeable into the tumour cells and pulsing the tumour site will increase the uptake of the drug from the systemic circulation into which the drug has previously been injected (Dev and Hofmann, 1996). In a study with six patients, electrochemotherapy was used for delivery of bleomycin to a total of 18 nodules. Three of the patients had malignant melanoma, two had basal cell carcinoma and one had metastatic adenocarcinoma. Eight 99 μs pulses of 1.3 kV/cm were administered to the tumours 5 to 15 min after intravenous administration of bleomycin to the patients. Square-wave pulses were delivered through electrodes placed on both sides of the protruding tumours. Five of the six patients responded positively to the treatment, with responses (partial to complete regression) seen in 13 of the 14 nodules. Some nodules were left untreated as positive controls. The patient who failed to respond had severe peripheral vascular disease, which may have compromised the delivery of drug to the tumour (Heller *et al.*, 1996). In order to avoid systemic drug delivery for localized therapy, bleomycin can potentially be injected directly into the tumours for electrochemotherapy. Studies in mice have shown that intratumoural injection of bleomycin in combination with electric pulses is effective and human studies are currently ongoing (Heller *et al.*, 1997).

The parallel plate electrodes used in the human study described earlier can treat tumours which are at the level of the skin but cannot be used on internal organs. Furthermore, they are difficult to fit and hold in place and the electric fields may not penetrate the deepest regions of cutaneous and subcutaneous tumours. Improvements in electrode design have thus been investigated. A needle array arrangement composed of six 28-gauge stainless steel acupuncture needles spaced at 60° intervals around a 1-cm diameter circle and extending 1 cm from the electrode body was found to be particularly useful. These needles can be inserted into the tumour to a depth of 0.5 cm. Each needle has an independent electrical connection so that a pulse sequence to energize individual needles can be developed. The design resulted in a 50 per cent improvement in response rate compared with the standard parallel plate electrodes (Gilbert *et al.*, 1997). Genetronics, Inc. (San Diego, CA, USA) has recently been granted a patent on its needle array applicator which will allow a physician to electroporate deep-seated tumours (*ID Weekly Highlights*, July 2, 1997, Current Drugs Ltd, London). The response rate of tumours which have only partially responded to single electrochemotherapy treatments can be improved by using multiple treatments (Jaroszeski *et al.*, 1996). Electrochemotherapy has also been investigated for treatment of human colorectal cell lines (Kambe *et al.*, 1996) and liver tumours (Jaroszeski *et al.*, 1997). It should be noted that direct electric current alone has also been used for treatment of tumours by electrotherapy (Miklavcic *et al.*, 1997), therefore appropriate controls are required for all studies.

6.9 Commercial development of electroporation

Electroporation can expand the scope of iontophoresis to deliver peptides in greater quantities (Potts *et al.*, 1997). More importantly, electroporation can expand the scope of transdermal delivery to larger molecules than can be delivered by iontophoresis. If electroincorporation can be commercialized to deliver particulates successfully into skin, then no molecular size limitations would exist to what can be delivered through the skin. In terms of progress towards commercial development of electroporation for transdermal delivery of drugs, the field is in its infancy compared with the progress made for commercial development of iontophoresis. The miniaturization of technology to develop wearable patches is currently not feasible for electroporation owing to the need for a capacitor. However, Genetronics has developed a palm-sized generator which can deliver pulses to a medication patch attached to the skin (Shaw, 1997). Patent activity by several universities and companies such as the Massachusetts Institute of Technology (Weaver *et al.*, 1991), Genetronics (Hofmann, 1995b, c) and others shows the developing interest in this area. Cygnus has a patent on the application of a driving force to assist the delivery of drugs through skin or tissue which has been electroporated. For example, iontophoresis can be used to keep the permeability of pulsed skin high over an extended period of time, and repeat pulses can be applied to keep the skin permeable (Bommannan *et al.*, 1996).

Unpublished data from the author's laboratory on several peptide and non-peptide drugs show that when a typical iontophoresis current (0.5 mA/cm^2) applied over 2–4 h is compared with electroporation (100–500 V; 1–10 m; 1–10 ppm) over 0.5 h or less, the former generally results in better delivery enhancement. However, the combined use of iontophoresis and electroporation often gives better results than iontophoresis alone. If iontophoresis and electroporation are compared at the same total charge delivered, electroporation appears to be better based on some literature reports, but the safety of these two protocols is not likely to be the same, with iontophoresis conditions being more acceptable. In a study which compared the transport efficiency of electroporation and iontophoresis, the transport numbers were in the same range and were a function of voltage and current. They did not show any dependence on pulse length, rate, energy, waveform or total charge transferred. The area fraction of skin available to transport was larger during high voltage pulsing than during iontophoresis (Prausnitz et al., 1996b). Unlike iontophoresis, very few human studies have been done with electroporation. In one recently reported, double-blind study, garlic juice was used as a simple model compound to demonstrate the feasibility of using electroporation for transdermal delivery. Six pulses (100 V, 10 ms) were delivered on the volar side of each arm, with some pressure applied during and after pulsing. The intensity of tongue sensation and taste was scored by the volunteers, and it was shown that pulses and some pressure were required to obtain a positive response (Zhang and Hofmann, 1997).

The electric fields developed in the tissue during pulsing will have to be characterized. While these fields have normally been modelled by computer technology, direct measurement of the field has also been reported since the conductivities in the body are very different and complicated (Cheng et al., 1996b). Thus, the use of electroporation pulses to permeabilize the skin and then followed by low current iontophoresis could be a very useful technique to achieve high flux levels of a drug without the skin irritation which would result if high iontophoresis current were used instead. Also, iontophoresis alone will not be able to deliver high molecular weight drugs, even if high current is used. The use of electroporation in combination with ultrasound has also been investigated and it was found that application of ultrasound reduces the threshold voltage required to facilitate transdermal drug transport in the presence of electric fields (Kost et al., 1996). A combination of ultrasound with what appears to be electroporation is already used in clinical medicine for physical therapy applications and instrumentation to achieve this objective is available (Intelect Model 700-C, Chattanooga Group, Inc., Hixon, TN, USA).

7
Electrically assisted delivery of gene-based drugs to skin

7.1 Introduction

Gene-based drugs and gene therapy are expected to revolutionize therapeutics in the coming years. However, the clinical application of these agents is limited by several problems, such as limited physical and chemical stability, or undesirable attributes for adequate absorption or distribution. Thus, as these macromolecules are made available, it will be essential to formulate these drugs into safe, stable and efficacious delivery systems. Iontophoresis and electroporation provide one viable opportunity for the delivery of these agents and this application will be discussed in this relatively brief chapter. Since work in this area is just starting, detailed background information is given. Iontophoretic delivery of the representative bases (uracil and adenine), nucleosides (uridine and adenosine) and nucleotides (AMP, ATP, GTP and imido-GTP) across mammalian skin *in vitro* has been demonstrated (van der Geest *et al.*, 1996a). For the delivery of gene-based drugs, the types of challenges faced may be somewhat different from those faced by conventional or peptide drugs. Charge and delivery considerations need to be carefully considered. The pK_a value of the phosphate backbone of oligonucleotides is very low (approximately 2.2), so that the molecules are negatively charged over a wide range of pH. Differences in pH, therefore, have little or no effect on the charge/mass ratio of the molecule and are not likely to affect iontophoretic delivery (Oldenburg *et al.*, 1995) from this point of consideration. However, other considerations are involved, such as the pH-dependent electro-osmotic flow and the type of backbone in the oligonucleotide, so that pH would still be a major factor (Brand and Iversen, 1996). Since oligonucleotides and DNA are negatively charged, their iontophoretic delivery would normally be under the cathode. Since electro-osmotic flow is from anode to cathode (see Section 2.6), it will hinder transport and this needs to be taken into consideration. This electro-osmotic flow will be pH dependent and also its overall contribution will be dependent on the size of the molecule being transported. In Section 5.4.1, it was described how the enzymatic activity of the skin presents a barrier to the delivery of peptides. The skin also contains nuclease and phosphatase activity, and oligonucleotides and DNA penetrating the skin are susceptible to the activity of these enzymes. Since many nucleases require bivalent cations for their activity, they can be inhibited by EDTA, while phosphatase activity can be inhibited by inorganic phosphate. Thus, an appropriate choice of buffer systems may be able to minimize degradation of oligonucleotides during transport through skin.

7.2 Delivery of oligonucleotides

Antisense oligonucleotides are short lengths (typically, 15–30 nucleotides long) of single-stranded RNA or DNA which have base sequences complementary to a specific gene or its mRNA. They can inhibit gene expression, primarily by binding to the target gene mRNA, and thus have the ability to block disease-causing genes selectively from producing disease-associated proteins. The first generation of antisense oligonucleotides, phosphothiolates or PS-oligonucleotides, have reached the stage of human clinical trials for many diseases, especially cancer and viral diseases (Scanlon et al., 1995; Agrawal, 1996). For infectious diseases, a gene essential to the survival of the organism is chosen, while for non-infectious diseases, a gene whose protein helps to cause or maintain the disease is chosen (Crooke et al., 1996). A second generation of antisense oligonucleotides is also currently being designed by modification of the oligonucleotide backbone and these are expected to have improved stability, safety and efficacy (Agrawal, 1996). Antisense technology offers the most straightforward approach to take advantage of the elucidation of genomic sequences and thus collaboration between antisense and genomic companies is developing, along with significant patent activity in this very promising area (Crooke et al., 1996). Conventional delivery routes such as oral delivery are not suitable for oligonucleotides owing to their large size, high negative charge, biological instability towards intra- and extracellular nucleases (Hudson et al., 1996), rapid in vivo plasma elimination kinetics, poor cellular uptake and ineffective delivery to target site. Some of these drawbacks may be overcome by the use of controlled-release delivery systems such as biodegradable polymer matrices or by using cationic lipids or liposomes which may enhance cellular uptake of oligonucleotides by their attraction to the negatively charged surfaces of most cells (Crooke, 1995; Lewis et al., 1995; Jong et al., 1997).

Transdermal delivery offers one possibility for the delivery of antisense oligonucleotides. Also, some diseases may be best treated via topical application of these agents. Phosphorothiolate oligonucleotides complementary to transforming growth factor-β (TGF-β) mRNA have been designed to eliminate scars, which can be caused by excessive collagen deposition due to overexpression of TGF-β in wounded skin. Partially modified forms of these oligonucleotides have been shown to penetrate normal and tape-stripped damaged rat skin during in vitro studies. Higher amounts permeated through damaged skin. For normal skin, the oligonucleotide did not penetrate through skin into the receptor medium, presumably owing to its high molecular weight (MW 8000) and polyanionic charge (Lee et al., 1996). Penetration enhancers have been investigated in vitro for the skin penetration and retention of a series of antisense methyl phosphonate oligonucleotides using hairless mouse or human cadaver skin. As the molecular weight of the oligonucleotide increased, the penetration rate was found to decrease, though the study was limited to an 18-mer (MW 5500) oligonucleotide as the largest size (Nolen et al., 1994).

Larger oligonucleotides may be delivered by iontophoresis and/or electroporation, which also offers a means to avoid the use of penetration enhancers. Iontophoretic in vitro delivery of oligonucleotides across excised, full-thickness, hairless mouse skin has been investigated. The flux was found to decrease with increasing size, with the 10-mer being transported at a rate about sixfold faster than the 30-mer and twofold faster than the 20-mer (Figure 7.1). The skin was first pretreated with a buffer containing EDTA and

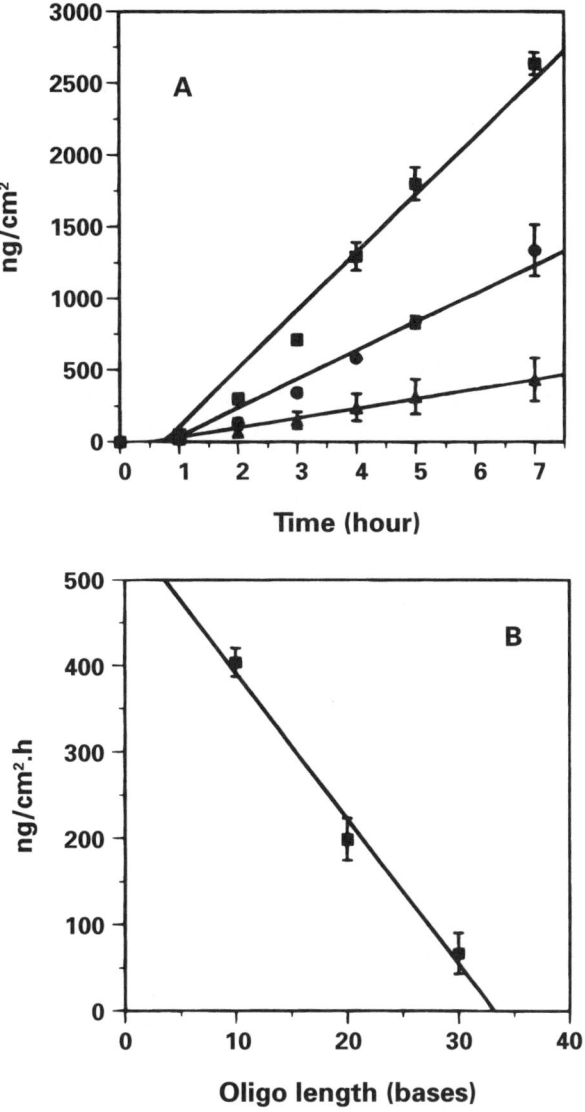

Figure 7.1 Effect of size on oligonucleotide iontophoresis. (A) Cumulative mass transport versus time for oligonucleotides of 10 (■), 20 (●) or 30 (▲) bases in length. Oligonucleotides were chemically synthesized to contain equal proportions of adenine, guanine, thymidine and cytosine at each position throughout the sequence. (B) Mean steady-state flux as a function of oligonucleotide length. A linear dependence is observed between the mean steady-state flux and oligonucleotide length. (Reprinted with permission from *Journal of Pharmaceutical Sciences*, **84** Oldenburg et al., Iontophoretic delivery of oligonucleotides across full thickness hairless mouse skin, pp. 915–921, Copyright 1995 American Chemical Society.)

K_3PO_4, which were also added to the transport buffer, as these are known to be effective inhibitors of nuclease and phosphatase activity, respectively. The buffer also contained 2.5 mg/ml of salmon testes DNA to prevent non-specific binding of the oligonucleotide to surfaces, during the studies on transdermal diffusion cells. EDTA or K_3PO_4 alone were not very effective, but when present together they resulted in the oligonucleotide being delivered mostly as an intact molecule in the receptor by iontophoresis. Dephosphorylation and/or degradation was observed only during transport of oligonucleotides through skin and was found to occur at a pH of 5.5 and below with virtually no degradation at pH 9.5. Incubation of the oligonucleotides in contact with the skin resulted in no degradation or dephosphorylation of the drug, suggesting that both nuclease and phosphatase activity is located within the tissue. A steady-state lag time of about 30–60 min was observed before the oligonucleotides appeared in the receptor chamber following the start of current, using a current density of 0.3 mA/cm^2 (Oldenburg et al., 1995). In contrast to this study, a small six-base sequence oligonucleotide, TAG-6 (MW 1927), was reported to cross hairless mouse skin with little or no degradation in an *in vitro* study. Cathodal iontophoretic delivery for 12 h using either the 5'-FITC or ^{35}S-labelled oligonucleotide resulted in a substantial flux, with steady-state levels of 273 ± 65 or 285 ± 71 ng/cm^2 per h respectively, suggesting that FITC labelling does not alter TAG-6 transport (Brand and Iversen, 1996).

Electroporation has been shown to be useful in improving intracellular delivery of synthetic antisense oligonucleotides to suppress a target protein (Bergan et al., 1993, 1996). The electroporation technique has been successfully used to transport 15-mer (4.8 kDa) and 24-mer (7.0 kDa) antisense oligonucleotides through human skin *in vitro* with fluxes of 6.4 and 11.5 pmol/cm^2 per h, respectively. The transport of fluorescein-labelled oligonucleotides was determined by measuring the fluorescence of the receptor compartment solution with a spectrofluorimeter. To visualize sites of local transport, a 2 per cent agarose gel was prepared in the receptor and lack of staining at sites corresponding to sweat ducts and hair follicles indicated that the transappendageal route was not a primary pathway. Transport was found to increase significantly with transdermal voltages greater than about 70 V, but then it formed a plateau soon afterwards. Transport also increased as the pulse length increased from 1.1 to 2.2 ms (Zewert et al., 1995). In another study, 15-mer phosphodiesters were delivered to hairless rat skin by electroporation. It was shown that the 3' protected phosphodiesters were better for topical delivery to the skin as a result of better stability against skin nucleases than phosphorothiolate oligonucleotides (Regnier et al., 1997).

7.3 Delivery of genes

In the past, most protein drugs were isolated from natural sources and then administered to patients after purification. With the advent of recombinant DNA technology, it became possible to produce human proteins in commercially viable quantities using organisms such as *Escherichia coli* or using mammalian cell lines. The next major advance in therapeutics is attempting to deliver human genes directly to the patient instead of transferring the gene first to an organism such as *E. coli* and then isolating and purifying the recombinant protein for administration to the patient. Thus, gene therapy may be considered a novel form of drug delivery in which the machinery of the patient's cells is used to produce a therapeutic agent to treat its own disease (Blau and Springer, 1995) and has various potential applications, including the cure or control of cancer (Cooper, 1996). *In vivo* gene therapy is based on introducing the genetic material directly into the patient, while *ex vivo* gene therapy involves removing tumour cells or lymphocytes from the patient and returning them after genetic modification via gene transfer. Though the methodology of gene transfer has been developed over many years, its first application to a human subject was in 1990 when a 4-year-old girl with adenosine deaminase (ADA) deficiency received an infusion of autologous T cells into which a normal ADA gene had been inserted (Stewart, 1995). The introduction of DNA into cells, called transfection, allows cells to produce proteins continuously, thus avoiding potential toxic effects which may result from high-dose bolus delivery of recombinant proteins. Delivery of proteins is also faced with the problem of a short half-life, thus requiring repeated administrations. Furthermore, proteins produced by transfected eukaryotic cells will undergo post-translational modifications such as glycosylation, unlike those produced in *E. coli* (Wicks, 1995).

The ability to target genes to the various layers, cell types and appendages of the skin is of interest for the subject matter of this book and could be used to correct skin disorders including inherited skin disorders, wound healing and those of ageing such as wrinkling (Zhang *et al.*, 1996; Trainer and Alexander, 1997). Several melanoma and non-melanoma skin cancers also have their origin in mutations caused by a variety of risk factors including ultraviolet radiation (Kanjilal *et al.*, 1995; Camplejohn, 1996) and may perhaps be treated by gene therapy. The involvement of an imbalance in the alleles on mouse chromosome 7, including the *H-ras* gene, has been implicated in mouse skin tumours (Portella *et al.*, 1994). Similarly, the damage induced in skin by ultraviolet light involves DNA damage (Qin *et al.*, 1995; Bennett *et al.*, 1996) and might possibly be treated by gene therapy. Topical diseases and conditions such as non-melanoma skin cancers, damage by ultraviolet light, lupus vulgaris, fungal infections, warts and herpes simplex virus infections can benefit from electroporation of nucleotide-based or other drugs. The skin is also an attractive target tissue owing to its accessibility.

While most of the work in the field of gene therapy has focused on viral gene transfer (i.e., using viral vectors), there is a need to develop non-viral approaches as these would offer several advantages. The primary advantage is that non-viral gene therapy will eliminate the risk of generating novel infectious agents. Also, non-viral methods can introduce large molecules of DNA into the cell, unlike viral systems which are limited by the carrying capacity of the virus. In addition, non-viral methods generally are associated with a lower risk of immunogenicity than viral methods (Slavikova and Massouridou, 1995; Treco and Selden, 1995). Liposomes and lipids have been extensively investigated as carriers for gene delivery. Cationic liposomes or lipids can complex with DNA, RNA or short-stranded antisense sequences and results have been very promising (Alexander and Akhurst, 1995; Lasic and Templeton, 1996; Wheeler *et al.*, 1996; Zelphati and Szoka, 1996; Bally *et al.*, 1997; Mahato *et al.*, 1997). A DNA/liposome complex applied *in vivo* to mouse skin was found to penetrate the skin rapidly and exhibit its β-galactosidase gene expression in epidermis, dermis and hair follicles. Expression was seen as early as 6 h after application and was high for 24–48 h, but reduced significantly by 7 days following application (Alexander and Akhurst, 1995). Similarly, even synthetic cationic polymers have been investigated as a vector for the design of *in vivo* and *ex vivo* gene transfection systems (Cherng *et al.*, 1996).

Electroporation is another technique which offers a non-viral method for gene therapy. Its general applications for transfection will be briefly discussed, followed by its specific application for delivery through skin. Electroporation, sometimes also called 'electropermeabilization', is a term used for the dramatic observations on changes in membranes of cells or artificial planar bilayer membranes which occur when large transmembrane voltages are applied. The changes in the membrane involve structural rearrangement and conductance changes leading to temporary loss of the semipermeability of cell membranes and suggesting formation of pores. These changes can lead to ion leakage, escape of metabolites and increased uptake by cells of drugs and DNA (Tsong, 1991; Weaver, 1993). As stated,

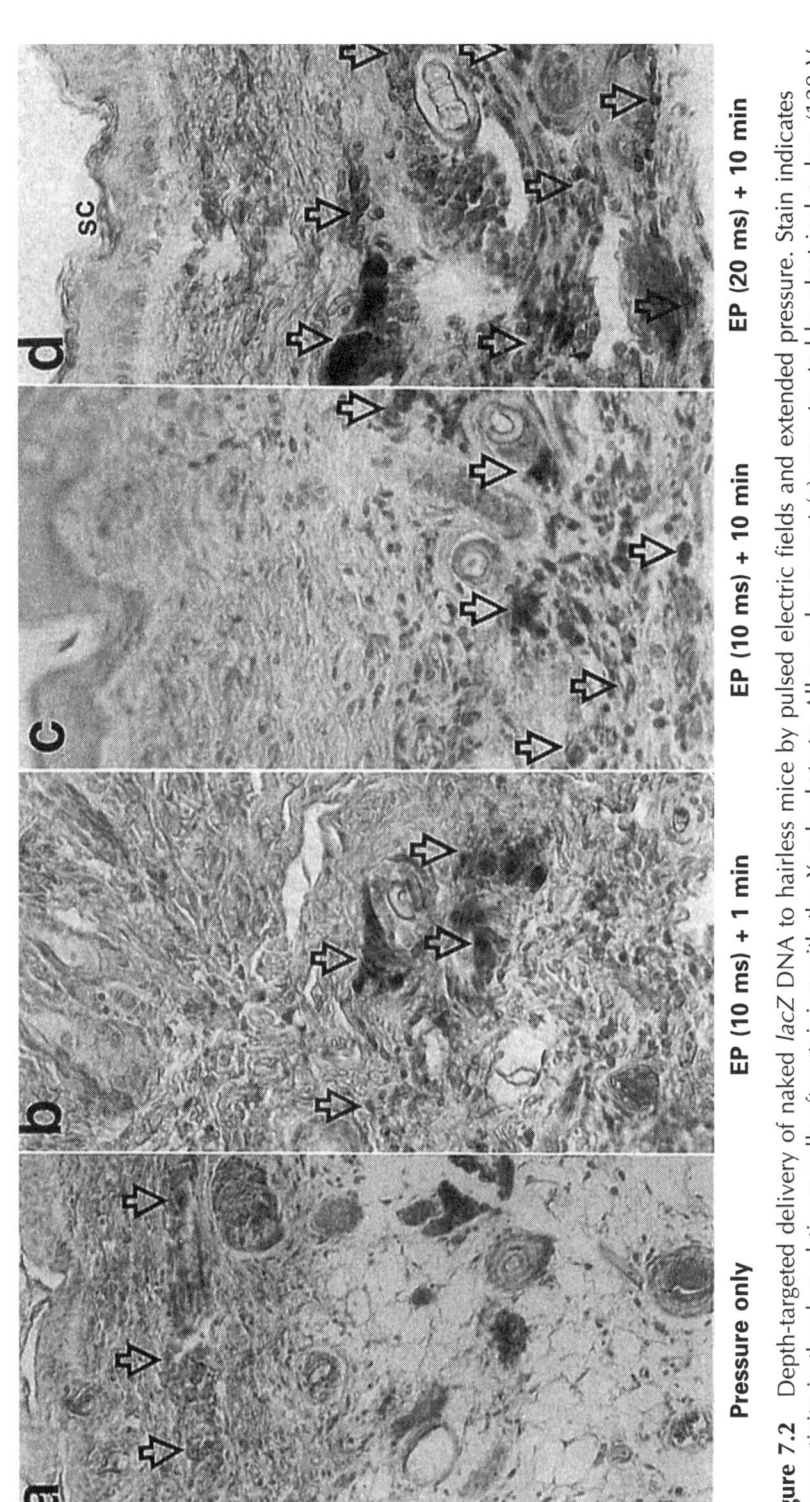

Figure 7.2 Depth-targeted delivery of naked *lacZ* DNA to hairless mice by pulsed electric fields and extended pressure. Stain indicates gene activity in the dermal tissue cells after staining with the X-gal substrate. All samples except (a) were treated by electrical pulses (120 V, 3 pulses). The pulse length and the duration of pressure were variable. (a) Control with calliper on the skin for 10 min without electrical pulses. (b) Pulse length 10 ms and pressure for 1 min during the pulses. (c) Pulse length 10 ms and pressure for 10 min after pulsing. (d) Pulse length 20 ms and pressure for 10 min after pulsing. Note the depth-targeted efficient gene delivery and expression in the skin (b–d). Note that the maximum depth of *lacZ* gene expression was much greater with extended pressure for 10 min after pulsing (c, d) compared with 1 min during the pulses (b). The number of transfected cells was more significant with a 20 ms pulse length than one of 10 ms. In the control (a), only light staining was found around hair follicles and only in the very upper layers of the skin. Light microscopy. Original magnification: × 125 (a), × 250 (b–d). Pressure only: gene expression was found only in hair follicles. (Reprinted from *Biochemical and Biophysical Research Communications*, **220** Zhang et al., Depth-targeted efficient gene delivery and expression in the skin by pulsed electric fields: An approach to gene therapy of skin aging and other diseases, pp. 633–636, Copyright 1996 with kind permission of

electroporation offers a non-viral method for gene therapy. It offers a widely investigated, simple and rapid technique that can also be used on non-replicating cells, unlike some other gene transfer methods (Matthews et al., 1995). The disadvantages of electroporation for gene therapy include low efficiency of stable transfection, the possibility of damage to a portion of the target population leading to cell death, and the possibility of integration of the transgene in concatameric form (Matthews et al., 1995; Dev and Hofmann, 1996). Transfer of DNA into cells by electroporation is a multiphase process and will be facilitated by the adsorption of DNA on the cell surface. Its subsequent transfer through the electroporated membrane will also be controlled by its polyelectrolyte nature and by the cell surface charges (Neumann, 1996).

Introduction of plasmid DNA into bacteria represents an important application of electroporation (Rosey et al., 1995; Li and Kuramitsu, 1996; Yang et al., 1996). Contact between plasmid DNA and bacteria must be present during the pulse and the proportion of electrotransformed cells is dependent on DNA concentration (Rittich and Spanova, 1996). Other factors determining the electrotransformation of bacteria by plasmid DNA have been investigated, including the use of electroporation buffer (Li et al., 1995), temperature (Wards and Collins, 1996), physical factors (Rittich et al., 1996), molecular form of DNA (Kimoto and Taketo, 1996) and other factors (Berthier et al., 1996). It has been reported that the antibiotic kanamycin can cause transient changes in the integrity of bacterial outer membrane, allowing the bacteria to become more electrocompetent (Mack and Titball, 1996). Electroporation has also been widely used to transform mammalian cells with plasmids (Tatsuka et al., 1995; Melkonyan et al., 1996) and has even been used to transform insect embryos (Leopold et al., 1996), plant embryos (Hansch et al., 1996), plant protoplasts (Wintz and Dietrich, 1996) and yeast cells (Neumann et al., 1996). The use of electroporation for gene therapy of skin has been shown in an *in vivo* study where *lacZ* DNA was delivered to hairless mice using three exponential decay pulses of amplitude 120 V and pulse length of 10–20 ms. After pulsing, some pressure was applied. The skin was disinfected with 70 per cent isopropyl alcohol before harvesting for X-gal staining 3 days after application of the *lacZ* DNA. The expression of the *lacZ* gene as indicated by blue staining was observed extensively in the dermis including the hair follicles (Figure 7.2). The driving force for this depth-targeted gene delivery was claimed to be controlled pressure after creating pores in the stratum corneum by pulsed electric fields (Zhang et al., 1996, 1997).

7.4 DNA vaccines

DNA vaccines, or DNA-mediated immunization, is based on the use of direct inoculation of plasmid DNA (pDNA) to raise immune responses by expression of an antigenic protein directly within the transfected cells. Using cloned genes in the form of plasmids allows one to introduce only the genes required for antigen production. In addition to providing the coding sequences for the antigenic protein, the plasmid also serves as the physical vector carrying the genes. This innovative approach to immunization holds true promise for the development of needed vaccines (Robinson et al., 1996). The promise of DNA vaccines was first shown in 1993 and since then has been demonstrated by several scientists, as reviewed recently (Donnelly et al., 1997), and includes the possibility of vaccination against genital infections (Donnelly et al., 1996), hepatitis B virus (Davis, 1996), murine cytomegalovirus (Armas et al., 1996), rotavirus (Herrmann et al., 1996), tuberculosis (Tascon et al., 1996) and HIV (Wang et al., 1995; Lu et al., 1996; Shiver *et al.*, 1996a, b). A strong immune response involving both the humoral and cellular arms of the immune system is involved. DNA vaccines are easier to manufacture than an inactivated pathogen or recombinant protein vaccine. Furthermore, the DNA is very stable even under high temperature conditions so that storage, transport and distribution will be easy.

Currently, intramuscular injection is commonly used, but it is now being realized that the skin may also provide a viable route for administration. The skin and mucous membranes are the anatomical sites where most exogenous antigens are commonly encountered. Following inoculation of DNA into skin, keratinocytes represent the predominant antigen-expressing cells. Though transfected keratinocytes will be lost in a few days due to the normal sloughing of the epidermis, it seems that relatively short-term expression of antigen is sufficient to raise long-term immune responses. However, expression of the gene in the DNA vaccine must be high because the DNA does not replicate in the mammalian cell. Thus, all expression comes from the small amount of DNA which is internalized by the cells. The Langerhans cells of the skin carry the antigen from the skin to the draining lymph nodes. These antigen-loaded Langerhans cells are potent activators of T lymphocytes, and a type of lymphocytes called epidermal lymphocytes play a key role in skin immunity (Raz *et al.*, 1994). pDNA may be administered 'naked' (dissolved in saline), complexed with lipids or dried on the surface of particles. Currently, the delivery of pDNA into skin cells is typically accomplished by direct intradermal injection or particle bombardment. Gene gun technology has been used for particle-mediated gene transfer to various organs, using various animal species (Yang and Sun, 1995). DNA molecules can be quantitatively coated onto microscopic gold particles (~ 1 to 3 µm) which can then be delivered into target tissue using a gene gun. Particle bombardment technology using a gene gun has been described in several publications (Yang *et al.*, 1990; Fuller *et al.*, 1996; Haynes *et al.*, 1996). Alternatively, a 'needleless' injection using a piston driven by compressed gas to deliver a thin stream of inoculum under high pressure can be used (Donnelly *et al.*, 1997). As discussed earlier, cationic liposomes can improve the delivery of DNA and thus have also been used for DNA vaccination (Gregoriadis *et al.*, 1997). Electrically assisted delivery, of course, forms the basis for this discussion and the feasibility of such delivery will be considered.

The skin-associated lymphoid tissue contains specialized cells that enhance immune responses. It has been shown that a single intradermal injection of less than 1 µg of specific DNA was sufficient to induce an immune response against the influenza virus. Gene expression in the epidermis and dermis was greater at 10 days than at 3 days after pDNA injection. As the epidermal cells (keratinocytes) are sloughed off, the expression remained mostly in the dermis at 30 days after injection (Raz *et al.*, 1994). While intradermal injection seems to work, the amount of DNA required for immunization can be reduced and immune responses can be improved if the DNA can be targeted directly to the cells, such as by using a gene gun to deliver DNA-coated gold beads to the epidermis. Once delivered to the cells, the DNA dissolves and can be expressed. In fact, it has been found that this route is more efficient than other parenteral or mucosal routes and needs as little as 0.4 µg of DNA, which is 250–2500 times less DNA than required with direct parenteral inoculation of purified DNA in saline. In this study, the expression was found to be transient with most being lost within 2–3 days due to the normal sloughing of the epidermis. However, as discussed, short-term expression of antigen is sufficient to raise long-term immune responses because epidermal Langerhans cells are capable of presenting transfected antigens to the T-helper component of the immune system, priming both T-helper and B-cell memory (Fynan *et al.*, 1993). Other reports on gene gun-mediated DNA vaccination via skin have also been published (Yang *et al.*, 1990; Benn *et al.*, 1996; Fuller *et al.*,

1996). Using a β-galactosidase marker gene, about 20 per cent of the skin epidermal cells were shown to express high levels of β-galactosidase activity (Yang *et al.*, 1990).

The technique of electroporation can also be used for DNA vaccination. Electroporation has been widely used to deliver DNA to single cell systems, but as discussed in Chapter 6, it has now been realized that electroporation of mammalian skin is feasible. One study has shown that *in vivo* electroporation can be used for the introduction of plasmid DNA into skin cells of mouse. A mixture of two supercoiled pDNAs was introduced subcutaneously to newborn mice and, after 10–60 min, the pleat of skin was exposed to two high-voltage pulses. The fibroblast primary cell cultures were obtained from the treated skin and after 2–3 weeks of selection, clones of stable transformed mouse fibroblasts were obtained (Titomirov *et al.*, 1991). In this study, DNA was first injected subcutaneously before the electroporation pulse was applied. It would be preferable to have a non-invasive means to introduce DNA into skin cells by electroporation. The application of electroporation assisted by controlled pressure to introduce DNA non-invasively into skin for gene therapy is discussed in Section 7.3.

8

Developmental issues and commercialization of wearable iontophoretic patches

8.1 Introduction

Iontophoresis is already approved in the USA, under an NDA, for the topical delivery of lidocaine and epinephrine for local analgesia (Iomed, Salt Lake City, UT), as discussed in Chapter 4. Devices available on the market for delivery of local anaesthetics and corticosteroids include Phoresor®II (Iomed), Empi® Dupel (Empi, Inc., St Paul, MN), Life-Tech® Iontophor (Houston, TX) and Henley Intl Dynaphor® (Houston, TX). In addition, devices for iontophoresis of pilocarpine for diagnosis of cystic fibrosis are on the market, and these include CF Indicator® (Scandipharm, Birmingham, AL) and the system based on the Webster sweat inducer with Pilogel® discs and a Macroduct® collector (Wescor, Inc., Utah, USA). These are discussed in Chapter 4. Commercialization of a wearable iontophoretic patch for self-use by the patient for systemic delivery of drugs will be discussed in this chapter. The inherent disadvantages of an iontophoretic patch over the marketed passive patches would include the complexity of the delivery system and cost considerations (Sarpotdar, 1991). Nevertheless, several companies are actively trying to commercialize miniature patch systems and are close to the market. A partial list includes the Alza Corporation (USA), Becton Dickinson (USA), Fournier (France), Hisamitsu (Japan) and Cygnus (USA). Several of these companies work on a contract basis for larger pharmaceutical companies. As a result, the number of pharmaceutical companies who are exploring iontophoresis technology is much larger than the number of companies with in-house capabilities for large-scale manufacture. For instance, some companies like Pharma Peptides (France), Sanofi Recherche (France) and Novartis (Europe/USA) have done some iontophoresis work. The collaboration will typically begin with an evaluation of technical feasibility and dermal tolerance of iontophoretic delivery for the drug. If initial studies are promising, then custom systems can be designed for exclusive marketing by the pharmaceutical partner (Becton Dickinson Transdermal Systems brochure on transdermal iontophoretic delivery, 1997). Dermion, a relatively new subsidiary of Iomed Inc., is also working on the development of wearable iontophoretic patches for contracts with pharmaceutical companies for their drugs. Iomed also recently bought the iontophoresis technology of the Elan Corporation (Ireland). There are hundreds of patents on iontophoresis between these companies and some individuals, with the majority

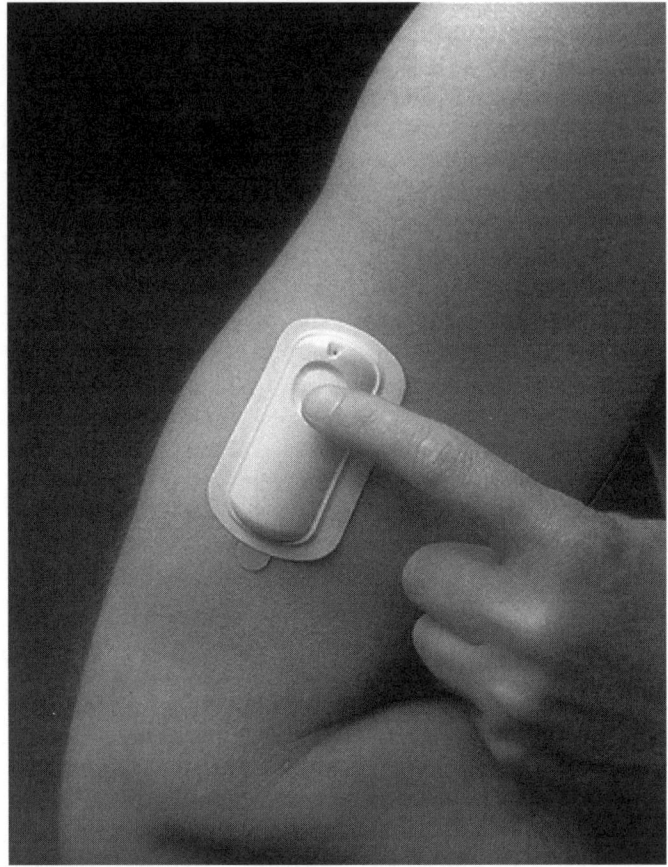

Figure 8.1 A patient demonstrating the use of the Alza's E-TRANS^{SM} on-demand electrotransport system. (By courtesy of Alza Corporation, Palo Alto, CA, USA.)

belonging to Alza and Becton Dickinson. Much of the activity in the area is proprietary and will unfold over the next few years. This chapter provides an overview of the information which is publicly known through presentations at scientific meetings, issued patents, annual reports, brochures, the Internet and published literature.

The Alza corporation has already invested at least about US$ 100 million in the electrotransport area (Felix Theeuwes, lecture given at the Annual meeting of American Association of Pharmaceutical Scientists, Seattle, October 27–31, 1996). Alza currently has an E-TRANS^{SM} fentanyl product under development with Janssen Pharmaceutica. Currently in phase III clinical trials, the product is an on-demand delivery system intended to allow a patient to manage acute pain by self-titrating the level of fentanyl administered according to his or her need. With this system, the onset for control of pain is almost instantaneous. Currently, transdermal fentanyl can only be used for chronic cancer pain. The development of an iontophoretic patch may allow treatment of acute pain, such as postoperative pain, since individualization and titration of dose will become feasible by changing electronic parameters. For details of published literature on transdermal/iontophoretic delivery of fentanyl, see Section 2.8.4. Alza is also working with its Therapeutic Discovery Corporation (TDC) to develop E-TRANS^{SM} electrotransport delivery of insulin (Alza 1996 Annual Report). A prototype of Alza's E-TRANS^{SM} electrotransport on-demand system is shown in Figure 8.1, which shows a patient demonstrating the use

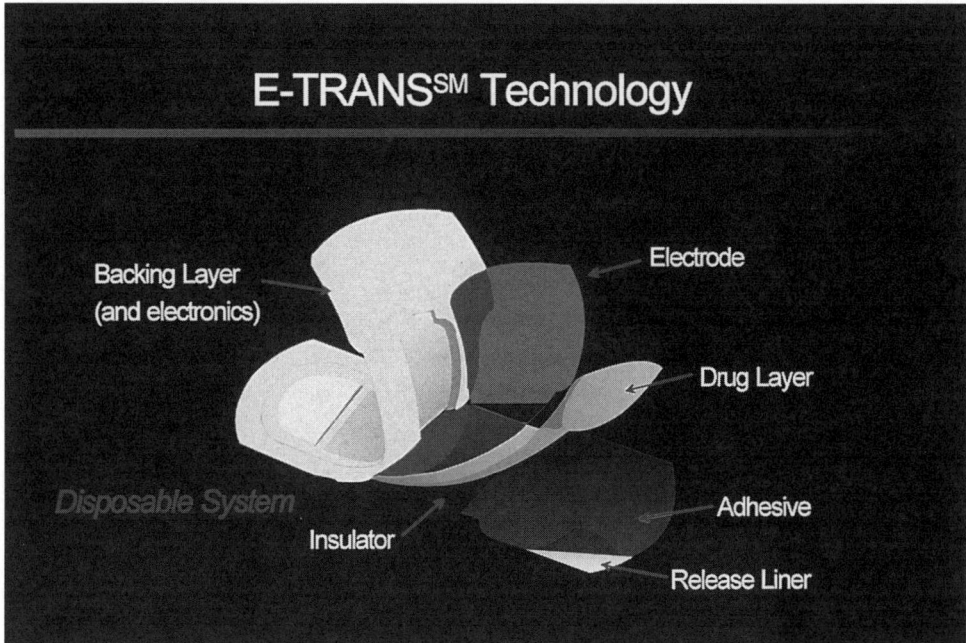

Figure 8.2 A schematic representation of the E-TRANS^{SM} system. (By courtesy of Alza Corporation, Palo Alto, CA, USA.)

of the system. A schematic representation of the system is shown in Figure 8.2. Both single use and reusable systems are in development. For prototypes of single use systems, the electronics and other components are thin and full integrated into the system, so that they are flexible and comfortable to wear. For reusable systems, a replaceable drug pad couples with the reusable controller. An on-demand system has a button to administer a bolus dose while the system continuously administers a baseline level of the drug. An LED display indicates whether the system is in baseline or bolus delivery mode. The size of these systems will vary from 5–50 cm^2, depending on scientific and marketing considerations. Since only small currents are employed, the system is safe for use on patients with pacemakers (*Electrotransport: A technology whose time has come*, brochure from Alza Corporation, 1993). These systems can deliver drug for periods from hours to days. Although comparable in size to Alza's passive transdermal systems, the company's electrotransport systems allow more drugs (including peptides) to be delivered by the transdermal route at therapeutic levels (E-TRANS^{SM} electrotransport technology leaflet, Alza Corporation, Palo Alto, CA).

A prototype of Becton Dickinson's iontophoresis patch is shown in Figure 8.3. These on-demand or patient-demand systems can provide flexible dosing patterns. Becton Dickinson is developing reusable controllers which can be connected and removed from the disposable housing. The controller monitors and controls the power supplied during use, thus permitting safer and more reliable operations. It also has the ability to detect the number of times the patch has been used, records the date and time of use, and its microprocessor can detect when drug supply is exhausted. Once the drug is exhausted, the controller can be rendered unusable to avoid abuse (Flower, 1996; Flower *et al.*, 1996). The reusable controller will allow costs to be kept down as the more expensive part of the device can be reused. In one concept, the controllers and reservoirs can be designed to fit together with a male/female interlock. A peel-away covering from the adhesive on the

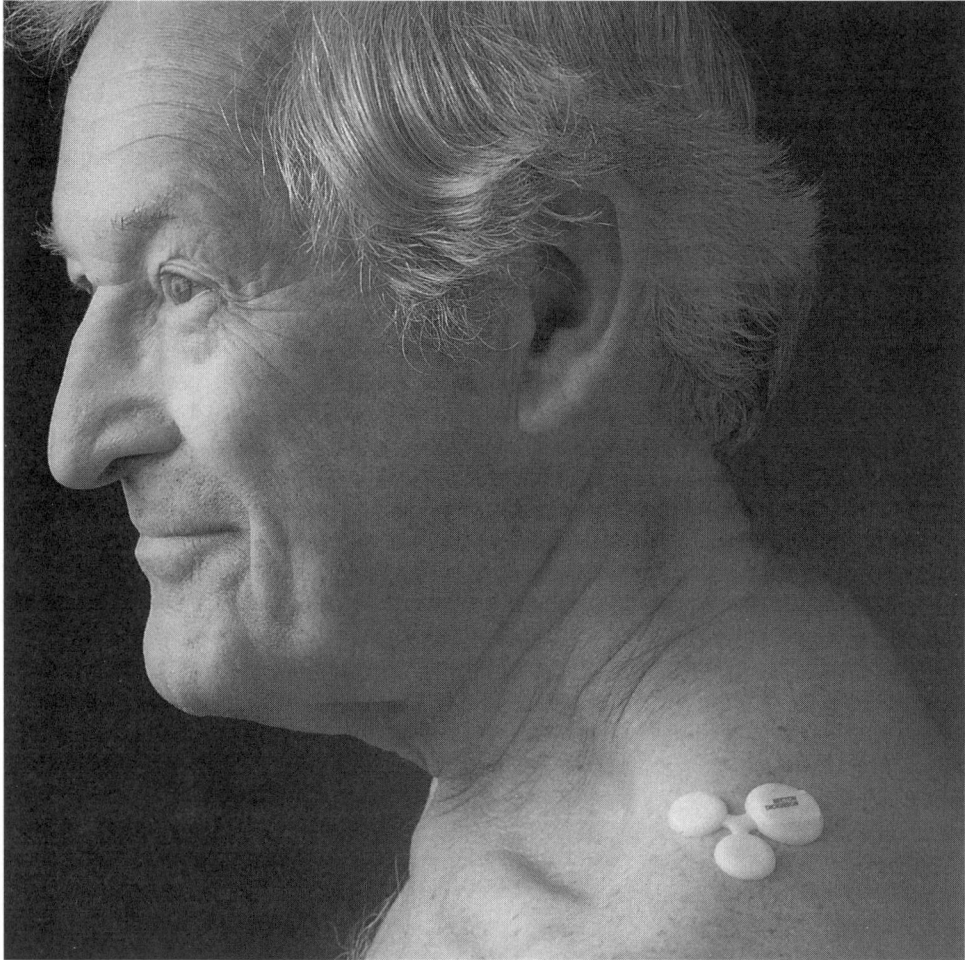

Figure 8.3 A patient demonstrating the use of a prototype of Becton Dickinson's iontophoresis patch. (By courtesy of Becton Dickinson, NJ, USA.)

controller side of the reservoir is removed and the reservoir is pressed against the controller to form the system assembly. Before use, a peel-away covering from the adhesive on the user side of the reservoir is removed and the patch pressed against the user.

Based on the principles of reverse iontophoresis (see Section 2.7), a glucose monitoring device (GlucoWatch™) is under development by Cygnus, Inc. (Redwood City, CA). The device consists of collection reservoirs, iontophoresis electrodes and sensor, and will be useful for continuous glucose monitoring, such as in neonates or subjects requiring frequent testing (Azimi *et al.*, 1996). Glucose is extracted from the body by electroosmosis induced by reverse iontophoresis and, using an electrochemical reaction linked with a sensor and a control module, blood glucose levels can be continuously monitored and displayed at the push of a button. The system can sound an alarm in the event of hypo- or hyperglycaemia. If approved, the device will be marketed and distributed by Becton Dickinson, which has partly sponsored the development costs for Cygnus. GlucoWatch is currently in development and clinical trials are expected to commence shortly (*ID Weekly Highlights*, July 2, 1997, Current Drugs Ltd, London). A 510 K application will be filed with the FDA for approval. Some of the problems that are being resolved include microbial growth in the saline pads, skin discomfort after 24 h of use, inability to measure glucose accurately during any significant sweating, and efforts towards miniaturization to wristwatch size. Compounds of interest for iontophoretic delivery in the patent literature include antimigraine drugs (De Beukelaar et al., 1995) and new derivatives of glucagon-like peptide 1 and insulinotrophin (Andrews et al., 1993). Insulin analogues have also been developed for potential use in iontophoretic delivery. These are prepared by substituting at least two of the residues in human insulin by Glu and/or Asp at selected positions. These analogues have a reduced tendency to associate and will thus overcome one of the obstacles to iontophoretic delivery of insulin (Brange, 1997). Other compounds of interest for iontophoretic delivery have been discussed throughout this book.

8.2 Human studies

Several human studies have been carried out for iontophoretic delivery. Many of these are done by companies for commercialization and have not been published yet. However, there are several reports in the published literature where iontophoresis has been performed in human subjects. While not all of these studies evaluated all the efficacy and safety endpoints, the results by and large have been promising. Clinically significant doses of fentanyl have been administered to humans with no adverse effects related to the delivery mode except for erythema at the location of the dispersive pad that resolved without treatment within 24 h (Ashburn et al., 1995). A discussion of iontophoresis of fentanyl can be found in Section 2.8.4 and comments about its commercial development are made later in this chapter. Iontophoretic delivery of leuprolide, an LHRH analogue, and a calcitonin analogue to human volunteers has also been reported and is discussed in Chapter 5, which is dedicated to the delivery of polypeptides by iontophoresis. The calcitonin analogue is the largest polypeptide so far reported to be delivered to man in controlled clinical trials. Octreotide, a bioactive peptide (MW ~1000), has been used in a placebo-controlled, dose-escalating study with three groups of eight healthy male volunteers. Steady-state levels were achieved quickly and increased with increasing magnitude of current. The plasma levels declined rapidly once the current was terminated (Figure 8.4) (Green, 1996b). Another study in four ethnic groups has also reported that a slight erythema observed at the active electrode disappeared within 24 h (Singh et al., 1994).

Iontophoretic delivery of alniditan, a 5 HT_{1D} agonist useful for treatment of migraine, has also been investigated in eight healthy volunteers. The molecule (MW 302) has pK_as of 8.3 and 11.5, and is ionized over a wide pH range, which hinders its passive transdermal permeation. An *in vitro* study was first performed on freshly excised hairless rat skin to optimize the factors affecting delivery. Based on these studies, the following conditions were used for the human studies: silver/silver chloride electrodes, ethanolamine buffer and pH 9.5. A dose of 0.5 mg of alniditan was successfully delivered within less than 1 h. An erythema detected at the anode lasted for about 48 h and could be linked to the high pH of the formulation, the current or the drug itself (Jadoul et al., 1996). In another study, significant amounts of salbutamol were absorbed in the two human subjects used 2 h after current application and plasma levels declined when the current was terminated (Bannon et al., 1988). Electrotransport of metoclopramide, an antiemetic, across the skin of healthy male subjects has also been reported rapidly to achieve and sustain reproducible blood levels (Phipps et al., 1994). Reverse iontophoresis studies for the extraction of glucose from the body have also been performed on humans as discussed in Section 2.7.

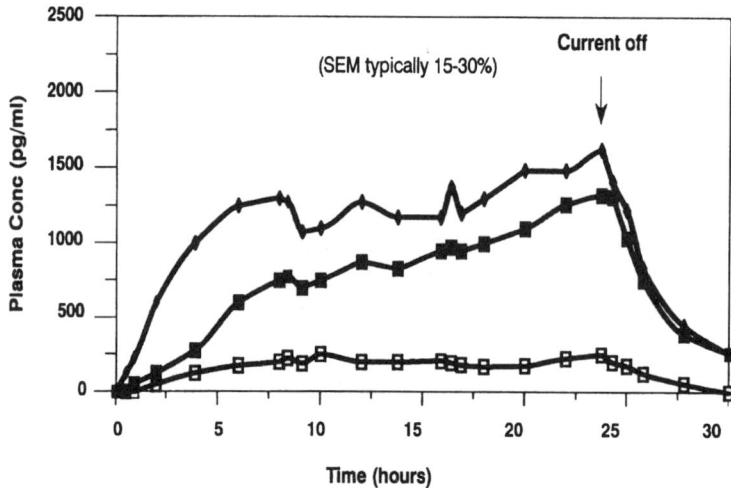

Figure 8.4 The effect of iontophoretic 24-h current levels on plasma concentration–time profiles of a peptide (MW ~1000) in healthy male volunteers ($n = 8$). □, 0.3 mA; ■, 0.6 mA; ◇, 1.2 mA. (Reprinted from *Journal of Controlled Release*, **41** Green, Iontophoretic delivery of peptide drugs, pp. 33–48, Copyright 1996 with kind permission from Elsevier Science – NL, Sara Burgerhartstraat 25, 1055 KV Amsterdam, The Netherlands.)

It should be realized that the skin can be quite variable between subjects or even at different sites on the same subject, with coefficients of variation as high as 50 per cent or more not being uncommon (Guy, 1996). The percutaneous absorption of ketoprofen in human subjects was similar when applied to the back or arm but lower when applied to the knee (Shah *et al.*, 1996). The effects of other variables such as age, race and gender on drug absorption through skin are less clear (Berti and Lipsky, 1995). The permeability coefficients of narcotic analgesics through human cadaver skin sites as diverse as the sole of the foot, chest, thigh and abdomen have been reported to be remarkably similar. Permeability through skin sections obtained from different cadavers varied four- to fivefold, but no trends in permeation as a function of age or gender were seen (Roy and Flynn, 1990). In general, ionic permeants produce more variable flux data than neutral permeants during passive transport through skin (Liu *et al.*, 1993). Ethnic differences may also influence absorption. In a comparison between American and Taiwanese smokers, the transdermal delivery of nicotine was higher for Taiwanese smokers. The differences between the ethnic groups were statistically significant for all patch sizes (Lin *et al.*, 1993b). In an iontophoresis study with four ethnic groups, the effect on the skin barrier function and irritation resulting from a 4-h application of current (0.2 mA/cm^2) on a 6.5 cm^2 area was evaluated. The ethnic groups were Caucasians, blacks, Hispanics and Asians, with 10 subjects in each group. Bioengineering skin instrumentation was used to monitor transepidermal water loss, capacitance, skin temperature and skin colour. The skin barrier function was not dramatically affected and racial differences observed were subtle, not major (Singh *et al.*, 1994). Racial differences in skin function in physiological and pathological conditions have been reviewed (Wester and Maibach, 1992; Berardesca and Maibach, 1996). An understanding of these differences can help to explain any observed differences in absorption, irritation, sensitization or erythematous reactions (Berardesca and Maibach, 1996). The incidence of skin irritation may also depend on the site of application (Berner and John, 1994).

8.3 Dose and bioavailability

The total dose delivered from an electric patch could potentially be much higher than from a passive patch. This is because iontophoresis uses an electric field to deliver the drug while a passive patch delivers drug by diffusion based on a concentration gradient. Using iontophoresis, more than 75 per cent of the drug can be administered by proper optimization of formulation, electrochemistry and device construction (Sage et al., 1995). However, typically the percentage of dose delivered in laboratory investigations is very low. During administration of LHRH, only 0.7 per cent of the dose was delivered systemically after a 5-h active period. In mass balance studies, 92 per cent of the dose could be accounted for, with the majority not even entering the skin (Table 8.1) (Heit et al., 1994b). Attention to electrode design during commercialization can greatly improve the dose delivered. If the drug forms a depot, then the ability to manipulate the drug once it has entered the skin will also affect the dose delivered to the circulation over a defined period of time.

Though iontophoresis will result in a faster input of drug, it has not been shown *in vivo* that iontophoresis can actually deliver a true bolus dose (Guy, 1996). Since the amount of drug delivered by iontophoresis depends on several factors, traditional dose–response studies cannot be performed (Sage and Riviere, 1992). Results may often be expressed as apparent bioavailability, which represents an underestimate of the absolute bioavailability which would result if the patches were used until all drug is exhausted. The apparent bioavailability of octreotide in the plasma of rabbits under optimal conditions (5 mg/ml at 150 $\mu A/cm^2$ for 8 h) was about 8 per cent (Lau et al., 1994). The use of the term 'bioavailability' for iontophoretic delivery can be confusing or misleading and the results should be interpreted only after considering the definition of bioavailability which was used. The following definition has been proposed:

$$\text{bioavailability} = \frac{\text{iontophoretic response/loaded mass}}{\text{reference response/loaded mass}}$$

Table 8.1 Mass balance tabulation of radiolabelled flaps (means and range of four flaps)

Location	Percentage of dose
Positive electrode	74.25 (62–82)
Negative electrode	0
Ring	15.5 (11–22)
Electrode assembly	89.75 (84–93)
Swab	1.1 (0.4–1.7)
Total not in skin	90.85 (85.7–94.2)
Stratum corneum (strip)	0.73 (0.2–1.5)
Underlying skin	0.35 (0.3–0.4)
Surrounding skin	0.25 (0.1–0.4)
Underlying subcutaneous tissue	0.04 (0.04)
Total in skin	1.37 (0.7–2.3)
Total recovery	92.2 (86.9–95.2)

(Reprinted from *Pharmaceutical Research*, **11** (7) Heit et al., Transdermal iontophoretic delivery of LHRH: Effect of repeated administration, pp. 1000–1003, Copyright 1994 with kind permission of Plenum Publishing Corporation.)

This definition relates to the pharmacodynamic bioavailability as the bottom line is to create a response in the patient (Sage *et al.*, 1995). Other definitions can also be used and pharmacokinetic bioavailability also needs to be calculated, but the perspective on definition and objectives should not be lost. The amount of current carried by any drug ion is a product of three major variables: the current efficiency, the electrode area and the current density. The current efficiency, in turn, depends on what fraction of the total current is being carried by the drug ion. This will be a factor of several variables such as the concentration of extraneous ions in the formulation.

Since smaller devices would be preferred by the patient and there is an upper limit to current density, methods to increase current efficiency such as by patch design will be critical to successful commercialization of this technology. The possible exception can be when the drug is relatively inexpensive and is not a narcotic (Sage and Riviere, 1992; Sage, 1993). A current density of 0.2 mA/cm^2 applied over 24 h is well tolerated and has been taken to clinical trials. Though the upper limit is generally considered to be 0.5 mA/cm^2, this current density can only be used for short periods of time. There have been very few systematic irritation studies carried out in man so far and these are currently in progress. Dermal irritation studies are initially done in rabbits using current density and duration conditions that are the same or similar to those to be used in humans. These rabbits studies are required by the FDA to initiate phase I studies in humans. If only transient erythema is observed, then human studies may be warranted. If more severe skin reactions are seen, the rabbit study will need to be repeated with reduced dosage of the drug. A saline patch study with current but no drug is also required. Sensitization studies in guinea pigs with an induction and challenge phase are also done. Sensitization or 'allergic contact dermatitis' involves an immunological response specific for the drug being used. The challenge phase occurs after a subsequent encounter of the sensitized subject with the inducing substance. Transdermal/topical literature in areas other than iontophoresis can provide useful information which may be applicable in many situations. An acute and cumulative irritation patch test in 151 volunteers has been described (Dreher *et al.*, 1996). Non-animal methods for testing skin sensitization are being developed (de Silva *et al.*, 1996) and should be evaluated for potential use.

8.4 Biological and safety issues

The immunology of the skin has been discussed in Section 1.2.3. It should be realized that skin is not a biologically inert membrane even though much of the transdermal literature treats it like one. Since iontophoretic transport takes place via an appendageal pathway, the immunology of the hair follicle may be important to understand and has been described in one chapter in a recent text on immunology of the skin (Paus, 1997). The skin will respond to any damage to its barrier by initiating a series of biochemical events designed to repair the damage, and the resulting level of irritation is reflective of the extent of perturbation caused by the enhancement mechanism (Guy, 1996). While several isolated literature reports exist on the adverse effects of application of current, it should be realized that many of these early studies were not well controlled in the sense that the electrochemical changes at electrodes were not controlled to prevent pH shifts or the pH of the drug formulation itself may have been low. However, transient erythema under the skin would be considered normal. In human studies carried out by the author in collaboration with coworkers (see Section 4.2.4), an erythema under the electrodes was noted that resolved within a few hours. A mild tingling sensation was reported as the current was being ramped up (Panus et al., 1997). Another report also describes a mild tingling sensation to the subjects as the current was ramped up to 0.25 mA/cm^2, with more tingling perceived under the anode. The sensation diminished with time of current application and disappeared in less than 30 min. The erythema observed under the electrodes resolved within 10–60 min of turning the current off. On occasion, a few, very small punctate lesions were observed that persisted for several days. Nevertheless, these side effects were similar to those that are observed with conventional transdermal drug delivery systems (Rao et al., 1995). The erythema associated with iontophoresis is most probably a result of non-specific irritation such as from an irritant drug or due to microscopic cellular damage at sites of high current density. This microscopic damage would release cytokine or prostaglandin resulting in local vasodilation. Direct electrical stimulation of C-fibres, a specific class of nociceptors, may also be responsible for sensations of tingling and pricking and development of erythema. C-fibres contain substance P and calcitonin gene-related peptide, the latter being a potent vasodilator which causes a localized erythema lasting over several hours. Substance P is involved in histamine-dependent weal and flare responses (Ledger, 1992).

Patent literature suggests that multivalent ions can reduce the sensation during iontophoresis (Phipps, 1995). Indeed, it has been shown that the mild sensation experienced by human volunteers when using NaCl for iontophoresis was reduced when CaCl$_2$ was used. The use of CaCl$_2$ led to a greater erythema but the rate of recovery of skin impedance was improved (Kalia and Guy, 1995). A more recent study has also shown that the type and concentration of electrolyte in the formulation affects the irritation response of the skin to iontophoresis (Anigbogu et al., 1997). In a comprehensive study on 30 pigs in vivo and 112 isolated perfused porcine skin flaps (IPPSFs), alterations in skin were seen following iontophoretic delivery of lidocaine hydrochloride. The change was characterized by light microscopy as the appearance of dark basophilic-staining nuclei oriented parallel to the stratum corneum in the stratum granulosum and spinosum layers. However, the dose-dependent, non-immune mediated, epidermal alteration was considered to have minimum toxicological significance (Monteiro-Riviere, 1990). Probably the worst electric burn would result if the electrode metal were allowed to touch the skin directly, as this would produce very high localized current density. It should be realized that any metal under a properly placed electrode pad will also cause shunting of current and can lead to a burn. Electric burns have been reported at the site of contact with a metal ring worn during iontophoresis (Reinauer et al., 1993). Even if the current density is uniformly applied to the skin, it should be realized that significant current density may flow through appendageal pathways as these are the primary pathways of transport. For the use of pilocarpine iontophoresis in a clinical setting for collection of sweat, it has been suggested that patients should be informed that the procedure carries a small risk of minor burns (Rattenbury and Worthy, 1996).

Safety features can easily be built into the circuitry of iontophoresis devices and several patents have been issued in this area. In order to prevent exposure of patients to excessive current, an intermediary storage device can be used that can use energy from the power supply and later transfer a predetermined safe level to the patient's skin to guard against failure of any component of the circuit (Haynes, 1994). Some patients may be sensitive to electricity, most probably as a result of psychosomatic illness (Liden, 1996), and electrically assisted delivery of drugs may be excluded in these individuals. While allergic contact dermatitis is rare in clinical applications of iontophoresis, it could develop because all drugs can be potentially allergenic and the iontophoresis patch has a relatively complex construction (Ledger, 1992). A recent case report describes a cutaneous allergic reaction to iontophoresis of 5-fluorouracil, which was reactivated at a distant site upon a subsequent treatment with iontophoresis (Anderson et al., 1997). Another potential concern may be dose dumping. However, solid-state circuits are quite reliable and can be set to avoid the discharge of excessive current in the remote chance that a circuit failure should happen. Similarly, if the current stops for some reason, means to alert the patient can be built into the device. In the event of an adverse reaction, the device can just be removed since the remaining dose resides in the patch outside the body (Sage, 1993).

8.5 Skin damage and its measurement

8.5.1 Skin damage

The electrical resistance of skin decreases with time during iontophoresis, the effect being larger at higher current densities. This decreased skin resistance often reflects increased permeability of the skin due to changes in the skin caused by current flow. For excised skin, such changes are partially irreversible and the resulting permeability change has been termed as the damage factor in the literature. The postiontophoresis passive permeability is thus generally higher than the passive permeability before iontophoresis. In one study, the passive flux measured after iontophoresis was about a factor of 10 greater than the corresponding flux measured before the skin was exposed to electric current (Pikal and Shah, 1990b). Similarly, the passive permeability of water and mannitol after 10 h of iontophoresis on hairless mouse skin was, respectively, 6 and 30 times greater than pretreatment values (Kim et al., 1993). The permeability increase or 'skin damage' factor may be evaluated by electrical conductivity (resistance) measurements. The time-dependent electrical resistance of porcine skin was found initially to drop quickly and then level off (Lin et al., 1997). Continuous membrane alteration has been shown to occur during in vitro iontophoresis of human epidermal membranes at high applied voltage drops (1000 mV), but does not occur when 250 mV is applied across the membrane. These changes appeared to reverse over time once the voltage drop was removed (Sims et al., 1992).

A recent study has investigated the effects of electric current on the fine structure of human stratum corneum lipids. The stratum corneum was exposed to pulsed constant current (0.013–13 mA/cm^2) for 1 h and then rapidly frozen and processed for freeze-fracture electron microscopy or subjected to X-ray diffraction analysis. A disordering of intercellular lipids was observed and was attributed to mutual repulsion following polarization of the lipid head group induced by the electric field. This change in the ordering of the intercellular lipid lamellae is responsible for the observed changes in skin resistance because both the resistance and capacitance properties of the skin are determined by the intercellular lipid lamellae. Other possible mechanisms which may also be involved include interactions of lipids with water or ions and displacement of structurally important ions such as Ca^{2+}. Any possible involvement of heat dissipation was shown to be unlikely by measuring the temperature close to the skin with thermosensors and showing that it remained constant throughout the experiment (Craane Van Hinsberg et al., 1997).

8.5.2 Measurement of skin damage

Bioengineering techniques such as transepidermal water loss (TEWL), electrical capacitance moist determination (ECM), laser-Doppler flowmetry (LDF) or impedance spectroscopy (IS) can be used to assess the damage to the skin during application of iontophoresis currents or electroporation pulses. Visual scoring can also be very useful. Passage of an iontophoresis current through human skin *in vivo* causes a significant reduction in the magnitude of skin impedance and it has been suggested that IS can complement TEWL as a technique for measuring *in vivo* skin permeability (Kalia and Guy, 1995). Actually, it seems that TEWL alone is inadequate to characterize fully the effect of iontophoresis on the barrier function of the skin. When the stratum corneum is physically damaged, both TEWL and IS are well correlated. However, when safe levels of iontophoresis current are applied, impedance is significantly reduced but TEWL is not affected. It would thus seem that intercellular lipid lamellae are not affected and there is no macroscopic perturbation of the stratum corneum. Thus, the passive transport of water is not affected. This lends support to a transappendageal pathway for iontophoresis (Kalia *et al.*, 1996).

IS, a non-invasive biophysical technique, can be used to study damage induced by chemical enhancers as well. IS has shown that enhancers increase the resistance of the skin as they open new penetration routes and increase ohmic resistance and capacitive properties, suggesting a rougher or more heterogeneous surface for excised human abdominal skin (Kontturi *et al.*, 1993). IS has also been used to investigate the extent to which the effect of iontophoresis on skin can be modulated by pretreatment with penetration enhancers (Kalia and Guy, 1997). A study using FTIR and DSC techniques has suggested that electrical treatment of the skin is less damaging than treatment of the skin with chemical penetration enhancers (Clancy *et al.*, 1994). The postiontophoresis recovery of hairless mouse skin following a 2-h exposure to 0.5 mA/cm^2 current has been investigated using IS and laser scanning confocal microscopic images of the permeability of calcein through pre-iontophoresed skin. The time for the recovery of skin impedance and permeability characteristics to pre-iontophoresis levels was found to be 18–24 h (Turner *et al.*, 1997). In a study of the percutaneous absorption of leuprolide, it was observed that a combination of a high level of ethanol and chemical enhancer was the main cause of skin irritation. In this study, rabbits were shaved at an abdominal site with a clipper and the skin was then allowed to recover for 24 h before applying the formulation via Hill Top Chamber patches (2.6 cm^2 each) (Lu *et al.*, 1992).

LDF is an optical technique which uses a laser beam to measure microcirculation, especially cutaneous blood flow, based on the backscattering of laser light by moving red blood cells. The cutaneous blood flow provides an indirect measure of the extent of irritation. It thus provides a simple non-invasive test which can provide continuous monitoring. Using LDF, it has been shown that novamide, a synthetic analogue of capsaicin, is much less irritating to humans than capsaicin upon topical application (Fang *et al.*, 1996d). The LDF technique was also used to assess the increased blood flow following iontophoresis of histamine in humans. When the flare disappeared, the level of LDF at the site of weal formation was still higher than the basal value (Thysman *et al.*, 1995a). Any irritation induced in humans can also be quantitatively and non-invasively measured based on changes in electrical impedance. The values obtained with impedance measurements seem to agree well with those obtained using TEWL and visual readings (Ollmar and Emtestam, 1992; Ollmar *et al.*, 1994). In a study with nine volunteers, the effect of iontophoresis on the stratum corneum barrier function and on development of erythema was measured using TEWL and LDF, respectively. Current was applied on the volar forearms of the subjects for 30 min. TEWL values measured after removal of the patch were about twofold higher than the baseline value irrespective of the presence or absence of current. This suggests that the effect of iontophoresis on the barrier to water loss is negligible and indistinguishable from that attributed to the occlusive effect resulting from the patch and/or formulation. The erythematous response, however, did differ from the controls and was 1.5–2.7 times higher at the anode than at the cathode. The erythematous response was considered moderate (van der Geest et al., 1996b). In addition to TEWL, other techniques such as capacitance, skin temperature, skin colour and a visual scoring system have also been used to assess any damage to skin barrier function following iontophoresis.

The first stage of a multistage study investigated the effect of 10 min of iontophoresis to validate the experimental methodology. The long-term goal of the study is to assess the safety of 24-h saline iontophoresis. Using 36 volunteers, current was applied at 0.1 mA/cm^2 on a 1-cm^2 patch or 0.2 mA/cm^2 on a 6.5-cm^2 patch and results were compared with controls. No damage to skin barrier function was seen, but a slight, subclinical, short-lasting, skin erythema was observed (Camel et al., 1996). Another study used LDF and TEWL techniques to study the safety of iontophoresis using six healthy female volunteers. A current of 0.1 or 0.2 mA/cm^2 was applied for 30 min. It was observed that iontophoresis increased cutaneous blood flow, with a more pronounced effect at higher current density. However, the increase was reversible within 1 h. TEWL values were not enhanced compared with controls (no iontophoresis), except for a small Joule heating effect at higher current density. The lipid structure of the skin was not affected and it was concluded that the drug delivery by iontophoresis should be safe (Thysman et al., 1995b). In this study, the TEWL and LDF measurements were the same at the anode and cathode sites. In contrast, another study with 26 healthy subjects suggests that the erythema was higher at the cathode as measured by LDF. This study was more of the type used in clinical medicine for physical therapy applications. Tap water iontophoresis was applied in a hydrogalvanic bath with separate cells for each arm. The current was sent through the body from one arm to the other, with the resulting current density being about 10 µA/cm^2 of body surface. A comparison of the microcirculation of the arms showed an increase of 120 per cent at the anode and of 700 per cent at the cathode, though the volunteers did not perceive any subjective difference between the electrode polarities (Berliner, 1997). Electrode details for this study are not provided and pH changes could at least partly be responsible for the observed effects.

The skin impedance has been measured in vivo in man for direct electrical evaluation of the effects of iontophoresis on tissue. At a current density of 0.5 mA/cm^2 applied for just a few minutes, the skin resistance dropped within seconds to essentially that of the electrolyte contained in the electrode chambers. The recovery of skin resistance was slow, with only 35 per cent of the pretreatment value being reached 4 h after the termination of current. These decreases are much faster than those typically seen with in vitro studies. The reasons for this discrepancy are not clear but may be related to damage to current-conducting pathways when skin is removed for in vitro studies, such as due to possible occlusion of sweat glands. For an equivalent current dose, short applications of high current were found to be more damaging to skin than longer applications at lower current density (Oh and Guy, 1995). The degree of erythema of skin can be quantified by a chromameter, as described in the literature (Kobayashi et al., 1997). X-ray diffraction technique has also been used and has shown that intercellular lipids are extensively reorganized by iontophoresis (Chesnoy et al., 1996).

8.6 Patch design and manufacture

Several, optimized, dose-efficient patches for peptide delivery have been discussed in the patent and published literature. In a patent granted to the author, a two-chamber system separated by a permselective membrane was proposed. The upper chamber had a pH-control mechanism by ion exchange resins, while the lower chamber could house a replaceable drug-loaded hydrogel device. The permselective membrane prevented the transport of drug into the upper chamber. The patch could be used for the delivery of conventional drugs or polypeptides such as vasopressin, calcitonin and insulin (Chien and Banga, 1993). In a somewhat similar design, the upper chamber contains halide ions and a silver electrode is used. Owing to problems related to use of such systems (see Section 3.4.6), it was suggested that the upper chamber may also be formulated with ion exchange resins or ion-selective membranes in order to prevent ions in the upper compartment entering the lower drug reservoir (Green, 1996b). Ion exchange materials that can be used include poly(acrylic acids), styrene and sulphonated divinylbenzene copolymers, polyamines and a ferrocyanide salt of a styrene divinyl benzene quaternary amine. The use of these materials, as well as the patch design, is described for fentanyl and other drugs (Gyory et al., 1995). Another patch described in the literature divides the two compartments by an ion exchange membrane. The chloride counterions cannot pass through the membrane and thus cannot compete for the current through the skin. The lower compartment contains only the drug formulation. Thus, the only ions involved are the drug ions and the ions from the skin, thereby maximizing the efficiency of the patch (Sage et al., 1995). In addition to providing increased dose efficiency, these designs also separate the electrode from the peptide. This will have the added advantage of preventing any possible reaction of the peptide at the electrode interface during iontophoresis. Special membranes such as those made from polysulphone, cellulose or fluororesin and treated with a cationic surfactant have low adsorptivity for peptides and have been used as an iontophoresis interface for their delivery (Higo et al., 1996). Membranes which are permeable to drug only when a voltage is applied have also been described (Cheng and Venkateshwaran, 1991).

The manufacture of an iontophoretic drug delivery patch is described in the patent literature. A well is created on a first laminate, filled with drug, and then a second laminate placed over the well and sealed to form a web from which individual patches are cut. The process is performed in an inert atmosphere which increases stability and shelf-life of the drug loaded into the iontophoretic patch (Broberg et al., 1996). A flexible conductive pathway and a flexible, non-conductive polymer film can be mated to form a flexible electronic circuit. This circuit can be disposable as it can be manufactured cheaply and can be coupled to other patch components without penetration (Haak et al., 1997). Optimization of the power supply to the patch can also be used to help with safety. For the palm-sized iontophoresis devices currently on the market for topical clinical applications (see Chapter 4), ramp-up of current is often built in to allow the patient to adjust to the feeling of current, thus minimizing the tingling sensation as the current is turned on. A microprocessor controls the electrode output current to rise exponentially from about zero to the selected treatment level. The current stays at the treatment level for the desired length of time and then falls exponentially to about zero (Fabian and Williams, 1995). A reversal of the polarity of electrodes during treatment has also been suggested to avoid tissue damage or eliminate undesirable skin sensations, in addition to allowing the use of delivery of multiple drugs with different polarity. Such an approach has been claimed to eliminate the need for buffering agents (Flower and Sage, 1995; Tapper, 1997). Possible degradation of the drug in direct contact with the electrodes has been discussed elsewhere in this book. This can be prevented by isolating the drug from the electrode with proper patch design. Degradation under the electric field itself is unlikely. The commonly used analytical method, capillary zone electrophoresis exposes peptides and proteins to much higher fields that those generated in an iontophoresis device (Sage et al., 1995).

8.7 Developmental and regulatory issues

First and foremost, there should be a clear therapeutic rationale for the selection of the drug candidate for iontophoretic delivery. The drug must be potent, with dosage preferably being only a few milligrams per day as this can be easily delivered. The drug should have a low or zero skin irritation or sensitization potential. Local irritation and the potential to induce an allergic response have been implicated as the factors which were often discovered late in the development path for transdermal patches and caused failure and high financial loss (Flynn and Stewart, 1988). The drug must also have the desired physicochemical properties to qualify as a viable candidate for iontophoretic delivery. The cost-effectiveness of the delivery system will depend on several factors including the cost of the drug itself. Commercialization of a wearable iontophoretic dosage form will require considerations of regulatory path, clinical protocols to prove safety and efficacy, and problems relating to product development. It is important to understand clearly the optimal dosing schedule and the feasibility of achieving it before proceeding too far along the developmental track (Cullander and Guy, 1992). Unlike the iontophoresis electrodes for topical application (see Chapter 4) which can be filled with virtually any drug, the design of a wearable patch will be specific to the drug. Thus, each drug will have its own iontophoresis system with a unique formulation and perhaps a unique design as well. Cyclic voltammetry studies need to be done to see how the drug may react to electrode surfaces or may degrade when the current is applied. The formulation must be carefully optimized considering all the factors which affect iontophoretic delivery. Excipients used should have poor current efficiencies so that they do not compete for the current with the drug.

The selection of the battery technology will be a critical part of a wearable system since miniaturized electronics must be used. The wafer form of primary cells such as lithium or mercury will last only a day or two while rechargeable cells may have unacceptable dimensions. A split battery technology in which air is consumed to energize the battery may be promising for the development of a wearable patch. In this system, the cathode is not consumed while the electrolyte and anode are a part of the drug reservoir which can be changed daily. With each change, a fresh battery is created. The system is small, light, inexpensive and efficient (Sage et al., 1995). An iontophoretic system could also be powered by an assembly of thermocouples (thermopile), which converts thermal energy into electrical energy. A thermopile has been demonstrated to deliver an ionic compound across human cadaver skin, with an efficiency of 3 $\mu g/\mu A$ per h, which is adequate to achieve therapeutic levels of a drug (Dinh et al., 1996).

For development of iontophoretic patches for all drugs, but especially for peptides, there will be several stability challenges. If the peptide does not have an adequate shelf-life in solution, it may have to be marketed as a dry powder to be reconstituted before use. The best way to achieve this objective would be via the use of hydrogels (see Section 3.5.2). The dry hydrogel can be activated by the user just before use (Green, 1996b). The patent literature describes several means to accomplish this objective. A burstable pouch can be used and ruptured just before use to provide the hydrating liquid. Alternatively, pulling a tab just before use can release the hydrating liquid from an aqueous reservoir which in turn can hydrate the dry hydrogel (Gyory and Perry, 1994). Another patent describes a variation where the end tab of a strip is pulled to cause progressive unsealing of the compartment to hydrate a matrix slowly (Beck et al., 1996). Polymers which are electronic conductors as a dry film may be very useful if they can be commercially developed for iontophoresis. Such polymers can be directly coated onto the electrodes and serve as drug reservoirs. In one study, a composite polymer film composed of cationic poly(N-methylpyrrole) (PMP^+) and anionic poly(styrene sulphonate) (PSS^-) was used. Reduction would lead to neutralization of the cationic backbone and incorporation of cations, such as protonated amine, dopamine. The PMP^+/PSS^- film on carbon was loaded by cathodic reduction at -0.6 V in a 0.1 M dopamine solution. Dopamine could then be released from this film in proportion to the current (Miller et al., 1987). The normal development and quality control considerations for a passive patch may be applicable to an iontophoretic patch as well, in addition to several tests which will be unique for the iontophoretic patch. The typical quality control, formulation, developmental and scale-up considerations for a passive patch have been described in the literature (Godbey, 1997; Van Buskirk et al., 1997; Vollmer and Cordes, 1997).

In vitro release studies may be required to assure batch-to-batch uniformity of product. These are similar to the studies that are required to assure batch-to-batch uniformity for topical and transdermal products. These studies may not predict the in vivo delivery and are only intended to detect any problems or variations in the product during routine production. For instance, crystallization of the drug from solution, its non-homogeneous distribution in polymer matrix or electronic malfunctions are likely to be detected by such studies. Synthetic membranes can be used for such studies. A discussion about the various synthetic membranes used in iontophoresis research can be found in Section 3.2.5. In fact, skin should be avoided for such studies as the high variability of data resulting from the skin factor may mask problems related to product performance. By the same logic, a rate-determining membrane is not required as the intention is to perform quality control on the patch, not the external release membrane. The membrane should be hydrophilic so that it allows the passage of current but it should not allow excessive movement of water. In addition to in vitro release studies, current profiles from the controller may be monitored by instrumentation such as oscilloscopes to detect any electronic malfunctions (Green, 1996a). A rotary disc cell approach to study iontophoretic release of drug from hydrogel vehicles has been described in the literature. A normal dissolution tester is modified to provide current to the rotating disc via the shaft and release across an artificial membrane into the vessel is monitored (Moll and Knoblauch, 1993).

Regulatory approval for an iontophoretic system will require an NDA submission, which should include a description of patch electronics in addition to information about the drug. The US regulatory pathways may differ depending on whether a prefilled iontophoretic drug delivery system or a reusable controller with a replaceable patch configuration is being developed (Green, 1996a). The most important factor in determining whether a drug delivery device is a medical device or a medicinal product is to establish whether the device forms an integral product with the drug. Based on the European Commission guidance document, single use, disposable iontophoresis devices incorporating a drug may be regulated as a medicinal product rather than a medical device (Donawa, 1997). The stability study should be not just for the drug but for other components as well, both alone and in combination. The possibility and implications of electronic malfunction and battery leakage should be addressed. The adhesive must adhere well to the skin or electric contact will be lost. During iontophoresis, the excipients may migrate into the skin in addition to the drug. This possibility needs to be addressed. Such co-iontophoresis of excipients may be desired in some situations, such as delivery of aprotinin to minimize degradation of peptides in skin (Chien and Banga, 1993), but in most instances it will not be desirable.

Rotation of application site may be required similar to that done for the currently marketed passive transdermal patches. If the electrode is placed back on the same area, it may generate more flux during the second active period, as was observed during iontophoretic delivery of LHRH. This may result because of saturation of skin binding sites during the first application so that the second application drives the drug directly into the vasculature or may even be causing desorption of the reservoir created during the first application. Alternatively, the barrier function of the skin may be compromised by the first application (Heit *et al.*, 1994b). Thus, the electrode should be placed on an unused portion of the skin for the second application. It is in the best interest of the handful of companies that have manufacturing expertise in the area of electrically assisted transdermal delivery to get together and propose common testing to regulatory authorities to self-regulate their production.

Bibliography

ABELL, E. and MORGAN, K., 1974, The treatment of idiopathic hyperhidrosis by glycopyrronium bromide and tap water iontophoresis, *Br. J. Dermatol.*, **91**, 87–91.

ABRAMSON, H.A. and GORIN, M.H., 1940, Skin reactions. IX. The electrophoretic demonstration of the patent pores of the living human skin; its relation to the charge of the skin, *J. Phys. Chem.*, **44**, 1094–1102.

AGRAWAL, S., 1996, Antisense oligonucleotides: Towards clinical trials, *Trends. Biotech.*, **14**, 376–387.

AGUILELLA, V., KONTTURI, K., MURTOMAKI, L. and RAMIREZ, P., 1994, Estimation of the pore size and charge density in human cadaver skin, *J. Control. Release*, **32**, 249–257.

AHMED, S., IMAI, T. and OTAGIRI, M., 1995, Stereoselective hydrolysis and penetration of propranolol prodrugs: *In vitro* evaluation using hairless mouse skin, *J. Pharm. Sci.*, **84**, 877–883.

AKHTER, S.A., BENNETT, S.L., WALLER, I.L. and BARRY, B.W., 1984, An automated diffusion apparatus for studying skin penetration, *Int. J. Pharm.*, **21**, 17–26.

AKINS, D.L., MEISENHEIMER, J.L. and DOBSON, R.L., 1987, Efficacy of the drionic unit in the treatment of hyperhidrosis, *J. Am. Acad. Dermatol.*, **16**, 828–832.

ALEXANDER, M.Y. and AKHURST, R.J., 1995, Liposome-mediated gene transfer and expression via the skin, *Hum. Mol. Genet.*, **4**, 2279–2285.

ALLEN, L.V., 1992, Dexamethasone iontophoresis, *US Pharmacist*, November, 86.

ANDERSON, L.L., WELCH, M.L. and GRABSKI, W.J., 1997, Allergic contact dermatitis and reactivation phenomenon from iontophoresis of 5-fluorouracil, *J. Am. Acad. Dermatol.*, **36**, 478–479.

ANDREWS, G.C., DAUMY, G.O., FRANCOEUR, M.L. and LARSON, E.R., 1993, New derivatives of glucagon-like peptide 1 and insulinotropin – used for enhancing insulin action in a mammal, particularly by iontophoresis, Pat. WO 9325579. Assignee: Pfizer; Scios.

ANIGBOGU, A., SINGH, P., LIU, P., DINH, S. and MAIBACH, H., 1997, Effects of iontophoresis on rabbit skin *in vivo*, *Pharm. Res.*, **14**, S-308.

ARIMA, H., ADACHI, H., IRIE, T. and UEKAMA, K., 1990, Improved drug delivery through the skin by hydrophilic β-cyclodextrins: Enhancement of anti-inflammatory effect of 4-biphenylacetic acid in rats, *Drug Invest.*, **2**, 155–161.

ARMAS, J.C.G., MORELLO, C.S., CRANMER, L.D. and SPECTOR, D.H., 1996, DNA immunization confers protection against murine cytomegalovirus infection, *J. Virol.*, **70**, 7921–7928.

ARVIDSSON, S.B., EKROTH, R.H., HANSBY, M.C., LINDHOLM, A.H. and WILLIAM-OLSSON, G., 1984, Painless venipuncture. A clinical trial of iontophoresis of lidocaine for venipuncture in blood donors, *Acta Anaesthesiol. Scand.*, **28**, 209–210.

ARVINTE, T., CUDD, A. and DRAKE, A.F., 1993, The structure and mechanism of formation of human calcitonin fibrils, *J. Biol. Chem.*, **268**, 6415–6422.

ASHBURN, M.A., STEPHEN, R.L., ACKERMAN, E., PETELENZ, T.J., HARE, B., PACE, N.L. and HOFMAN, A.A., 1992, Iontophoretic delivery of morphine for postoperative analgesia, *J. Pain Symptom Mgmt.*, **7**, 27–33.

ASHBURN, M.A., STREISAND, J., ZHANG, J., LOVE, G., ROWIN, M., NIU, S., KIEVIT, J.K., KROEP, J.R. and MERTENS, M.J., 1995, The iontophoresis of fentanyl citrate in humans, *Anesthesiology*, **82**, 1146–1153.

AUGUSTIN, C., FREI, V., PERRIER, E., HUC, A. and DAMOUR, O., 1997, A skin equivalent model for cosmetological trials: An *in vitro* efficacy study of a new biopeptide, *Skin Pharmacol.*, **10**, 63–70.

AULT, J.M., LUNTE, C.E., MELTZER, N.M. and RILEY, C.M., 1992, Microdialysis sampling for the investigation of dermal drug transport, *Pharm. Res.*, **9**, 1256–1261.

AUNGST, B.J., 1988, Nicotine skin penetration characteristics using aqueous and non-aqueous vehicles, anionic polymers, and silicone matrices, *Drug Dev. Ind. Pharm.*, **14**, 1481–1494.

AVITALL, B., 1992, Myocardial iontophoresis, U.S. Pat. 5087243. Assignee: Avitall B.

AZIMI, N.T., BHAYANI, B.V., CAO, M., LEE, R.K., LEUNG, L., PLANTE, P.J., TAMADA, J., TIERNEY, M.J. and VIJAYAKUMAR, P., 1996, Iontophoresis sampling appts for transdermal monitoring of target substance – has collection reservoir comprising ionically conductive hydrogel and ionically conductive solution, and two iontophoresis electrodes in contact with collection reservoirs in contact with subject's skin, Pat. WO 9600110. Assignee: Cygnus.

BAGNIEFSKI, T. and BURNETTE, R.R., 1990, A comparison of pulsed and continuous current iontophoresis, *J. Control. Release*, **11**, 113–122.

BAHARLOO, M., 1987, Iontophoresis and phonophoresis: A role for the pharmacist, *Hosp. Pharm.*, **22**, 730–731.

BALLY, M.B., ZHANG, Y.P., WONG, F.M.P., KONG, S., WASAN, E. and REIMER, D.L., 1997, Lipid/DNA complexes as an intermediate in the preparation of particles for gene transfer: An alternative to cationic liposome/DNA aggregates, *Adv. Drug Del. Rev.*, **24**, 275–290.

BANERJEE, P.S. and RITSCHEL, W.A., 1989a, Transdermal permeation of vasopressin. I. Influence of pH, concentration, shaving and surfactant on *in vitro* permeation, *Int. J. Pharm.*, **49**, 189–197.

BANERJEE, P.S. and RITSCHEL, W.A., 1989b, Transdermal permeation of vasopressin. II. Influence of azone on *in vitro* and *in vivo* permeation, *Int. J. Pharm.*, **49**, 199–204.

BANGA, A.K., 1996, *Therapeutic Peptides and Proteins: Formulation, processing, and delivery systems*. Technomic, Lancaster, PA, pp. 1–316.

BANGA, A.K. and CHIEN, Y.W., 1988, Iontophoretic delivery of drugs: fundamentals, developments and biomedical applications, *J. Control. Release*, **7**, 1–24.

1993a, Hydrogel-based iontotherapeutic delivery devices for transdermal delivery of peptide/protein drugs, *Pharm. Res.*, **10**, 697–702.

1993b, Dermal absorption of peptides and proteins, in: *Pharmaceutical Biotechnology. Vol. 4. Biological Barriers to Protein Delivery*, K.L. AUDUS and T.J. RAUB (eds). Plenum Press, New York, NY, pp. 179–197.

1993c, Characterization of *in vitro* transdermal iontophoretic delivery of insulin, *Drug Develop. Ind. Pharm.*, **19**, 2069–2087.

BANGA, A.K., KATAKAM, M. and MITRA, R., 1995a, Transdermal iontophoretic delivery and degradation of vasopressin across human cadaver skin, *Int. J. Pharm.*, **116**, 211–216.

BANGA, A.K., KULKARNI, S. and MITRA, R., 1995b, Selection of electrode material and polarity in the design of iontophoresis experiments, *Int. J. Pharm. Adv.*, **1**, 206–215.

BANNON, Y.B., CORISH, J. and CORRIGAN, O.I., 1987, Iontophoretic transport of model compounds from a gel matrix across a cellophane membrane, *Drug Develop. Ind. Pharm.*, **13**, 2617–2630.

BANNON, Y.B., CORISH, J., CORRIGAN, O.I. and MASTERSON, J.G., 1988, Iontophoretically induced transdermal delivery of salbutamol, *Drug Develop. Ind. Pharm.*, **14**, 2151–2166.

BANTA, C.A., 1994, A prospective, nonrandomized study of iontophoresis, wrist splinting, and antiinflammatory medication in the treatment of early-mild carpal tunnel syndrome, *J. Occup. Med.*, **36**, 166–168.

BARBEAU, D., 1995, Rational approaches for extended release of cardiovascular drugs, *Cont. Rel. Newsletter*, **12**, 6–7.
BARRY, B.W., 1987, Mode of action of penetration enhancers in human skin, *J. Control. Release*, **6**, 85–97.
1991, Lipid-protein-partitioning theory of skin penetration enhancement, *J. Control. Release*, **15**, 237–248.
BARTEK, M.J., LABUDDE, J.A. and MAIBACH, H.I., 1972, Skin permeability *in vivo*: Comparison in rat, rabbit, pig and man, *J. Invest. Dermatol.*, **58**, 114–123.
BARZA, M., PECKMAN, C. and BAUM, J., 1986, Transscleral iontophoresis of cefazolin, ticarcillin and gentamicin in the rabbit, *Ophthalmology*, **93**, 133–139.
1987, Transscleral iontophoresis of gentamicin in monkeys, *Invest. Ophthalmol. Vis. Sci.*, **28**, 1033–1036.
BAUER, H.H., AEBI, U., HANER, M., HERMANN, R., MULLER, M., ARVINTE, T. and MERKLE, H.P., 1995, Architecture and polymorphism of fibrillar supramolecular assemblies produced by *in vitro* aggregation of human calcitonin, *J. Str. Biol.*, **115**, 1–15.
BAYON, A.M.R. and GUY, R.H., 1996, Iontophoresis of nafarelin across human skin *in vitro*, *Pharm. Res.*, **13**, 798–800.
BAYON, A.M.R., CORISH, J. and CORRIGAN, O.I., 1993, *In vitro* passive and iontophoretically assisted transport of salbutamol sulphate across synthetic membranes, *Drug Develop. Ind. Pharm.*, **19**, 1169–1181.
BECK, J.E., LLOYD, L.B. and PETELENZ, T.J., 1996, Iontophoretic delivery device with integral hydrating assembly – has sealed liquid storage compartment with tab pulled for progressive unsealing of compartment to release liquid onto hydratable matrix, WO Pat. 9605884. Assignee: Iomed.
BELLANTONE, N.H., RIM, S., FRANCOEUR, M.L. and RASADI, B., 1986, Enhanced percutaneous absorption via iontophoresis I. Evaluation of an *in vitro* system and transport of model compounds, *Int. J. Pharm.*, **30**, 63–72.
BENN, S.I., WHITSITT, J.S., BROADLEY, K.N., NANNEY, L.B., PERKINS, D., HE, L., PATEL, M., MORGEN, J.R., SWAIN, W.F. and DAVIDSON, J.M., 1996, Particle-mediated gene transfer with transforming growth factor-beta 1 cDNAs enhances wound repair in rat skin, *J. Clin. Invest.*, **98**, 2894–2902.
BENNETT, P.V., GANGE, R.W., HACHAM, H., HEJMADI, V.S., MORAN, M., RAY, S. and SUTHERLAND, B.M., 1996, Isolation of high-molecular-length DNA from human skin, *Biotechniques*, **21**, 458–461.
BERARDESCA, E. and MAIBACH, H., 1996, Racial differences in skin pathophysiology, *J. Am. Acad. Dermatol.*, **34**, 667–672.
BERGAN, R., CONNELL, Y., FAHMY, B. and NECKERS, L., 1993, Electroporation enhances c-myc antisense oligodeoxynucleotide efficacy, *Nucleic Acids Res.*, **21**, 3567–3573.
BERGAN, R., HAKIM, F., SCHWARTZ, G.N., KYLE, E., CEPADA, R., SZABO, J.M., FOWLER, D., GRESS, R. and NECKERS, L., 1996, Electroporation of synthetic oligodeoxynucleotides: A novel technique for *ex vivo* bone marrow purging, *Blood*, **88**, 731–741.
BERLINER, M.N., 1997, Skin microcirculation during tapwater iontophoresis in humans: Cathode stimulates more than anode, *Microvascular. Res.*, **54**, 74–80.
BERNER, B. and JOHN, V.A., 1994, Pharmacokinetic characterization of transdermal delivery systems, *Clin. Pharmacokinet.*, **26**, 121–134.
BERTHIER, F., ZAGOREC, M., CHAMPOMIERVERGES, M., EHRLICH, S.D. and MORELDEVILLE, F., 1996, Efficient transformation of *Lactobacillus sake* by electroporation, *Microbiology*, **142**, 1273–1279.
BERTI, J.J. and LIPSKY, J.J., 1995, Transcutaneous drug delivery: A practical review, *Mayo Clin. Proc.*, **70**, 581–586.
BERTOLUCCI, L.E., 1982, Introduction of antiinflammatory drugs by iontophoresis: Double blind study, *J. Orthop. Sports Phys. Ther.*, **4**, 103–108.
BEZZANT, J.L., STEPHEN, R.L., PETELENZ, T.J. and JACOBSEN, S.C., 1988, Painless cauterization of spider veins with the use of iontophoretic local anesthesia, *J. Am. Acad. Dermatol.*, **19**, 869–875.

BHATIA, K.S. and SINGH, J., 1997, Percutaneous absorption of LHRH through porcine skin: Effect of N-methyl 2-pyrrolidone and isopropyl myristate, *Drug Develop. Ind. Pharm.*, **23**, 1111–1114.

BHATIA, K.S., GAO, S., FREEMAN, T.P. and SINGH, J., 1997a, Effect of penetration enhancers and iontophoresis on the ultrastructure and cholecystokinin-8 permeability through porcine skin, *J. Pharm. Sci.*, **86**, 1011–1015.

BHATIA, K.S., GAO, S. and SINGH, J., 1997b, Effect of penetration enhancers and iontophoresis on the FT-IR spectroscopy and LHRH permeability through porcine skin, *J. Control. Release.*, **47**, 81–89.

BI, R.C., DAUTER, Z., DODSON, E., DODSON, G., GIORDANO, F. and REYNOLDS, C., 1984, Insulin's structure as a modified and monomeric molecule, *Biopolymers*, **23**, 391–395.

BLAU, H.M. and SPRINGER, M.L., 1995, Gene therapy – A novel form of drug delivery, *N. Engl. J. Med.*, **333**, 1204–1207.

BODERKE, P., PONEC, M., BODDE, H.E. and MERKLE, H.P., 1997, Comparison of aminopeptidase activity in HaCaT cell culture sheets and in stripped human skin, *Proc. Int'l Symp. Control. Rel. Bioact. Mater.*, Controlled Release Society, Inc., 24. pp. 439–440.

BOERICKE, K., O'CONNELL, M.A., BOCK, C.R., GREEN, P., and DOWN, J.A., 1996, Iontophoretic delivery of human parathyroid hormone (1–34) in swine, *Proc. Int'l Symp. Control. Rel. Bioact. Mater.*, Controlled Release Society, Inc., 23. pp. 200–201.

BOGNER, R.H. and BANGA, A.K., 1994, Iontophoresis and phonophoresis, *US Pharmacist*, **19**, H10–H26.

BOMMANNAN, D.B., TAMADA, J., LEUNG, L. and POTTS, R.O., 1994, Effect of electroporation on transdermal iontophoretic delivery of luteinizing hormone releasing hormone (LHRH) *in vitro*, *Pharm. Res.*, **11**, 1809–1814.

BOMMANNAN, D.B., CHEN, T., POTTS, R.O. and WONG, O., 1996, Pulsed transport of substance through tissue using electrical pulse to cause electroporation – then using passive diffusion or iontophoresis with subsequent pulses applied after hours or even days, Pat. WO 9600111. Assignee: Cygnus.

BOND, J.R. and BARRY, B.W., 1988, Limitations of hairless mouse skin as a model for *in vitro* permeation studies through human skin: Hydration damage, *J. Invest. Dermatol.*, **90**, 486–489.

BORNMYR, S., SVENSSON, H., LILJA, B. and SUNDKVIST, G., 1997, Skin temperature changes and changes in skin blood flow monitored with laser Doppler flowmetry and imaging: A methodological study in normal humans, *Clin. Physiol.*, **17**, 71–81.

BOS, J.D., 1997a, The skin as an organ of immunity, *Clin. Exp. Immunol.*, **107**, 3–5.

BOS, J.D., 1997b, *Skin Immune System: Cutaneous immunology and clinical immunodermatology.* CRC Press, Boca Raton, pp. 1–719.

BOSMAN, I.J., LAWANT, A.L., AVEGAART, S.R., ENSING, K. and DE ZEEUW, R.A., 1996, Novel diffusion cell for *in vitro* transdermal permeation, compatible with automated dynamic sampling, *J. Pharmaceut. Biomed. Anal.*, **14**, 1015–1023.

BOXHALL, M. and FROST, J., 1984, Iontophoresis and herpes labialis, *Med. J. Aust.*, **140**, 686–687.

BOXTEL, A.V., 1977, Skin resistance during square-wave electrical pulses of 1 to 10 mA, *Med. Biol. Eng. Comput.*, **15**, 679–687.

BRAND, R.M. and GUY, R.H., 1995, Iontophoresis of nicotine *in vitro*: Pulsatile drug delivery across the skin? *J. Control. Release*, **33**, 285–292.

BRAND, R.M. and IVERSEN, P.L., 1996, Iontophoretic delivery of a telomeric oligonucleotide, *Pharm. Res.*, **13**, 851–854.

BRAND, R.M., DUENSING, G. and HAMEL, F.G., 1997, Iontophoretic delivery of an insulin-mimetic peroxovanadium compound, *Int. J. Pharm.*, **146**, 115–122.

BRANGE, J.J.V., 1997, Transdermal insulin, U.S. Pat. 5597796. Assignee: Novo Nordisk.

BROBERG, B.F.J., CLARK, R.J. and GREEN, P.G., 1996, Forming and packaging of iontophoretic drug delivery patches – in a continuous process in an inert atmosphere to provide increased shelf-life, Pat. WO 9610398. Assignee: Becton Dickinson.

BROUGH, K.M., ANDERSON, D.M., LOVE, J. and OVERMAN, P.R., 1985, The effectiveness of iontophoresis in reducing dentin hypersensitivity, *J. Am. Dent. Assoc.*, **111**, 761–765.

BROWN, L. and LANGER, R., 1988, Transdermal delivery of drugs, *Ann. Rev. Med.*, **39**, 221–229.

BURNETTE, R.R. and BAGNIEFSKI, T.M., 1988, Influence of constant current iontophoresis on the impedance and passive Na^+ permeability of excised nude mouse skin, *J. Pharm. Sci.*, **77**, 492–497.

BURNETTE, R.R. and MARRERO, D., 1986, Comparison between the iontophoretic and passive transport of thyrotropin releasing hormone across excised nude mouse skin, *J. Pharm. Sci.*, **75**, 738–743.

BURNETTE, R.R. and ONGPIPATTANAKUL, B., 1987, Characterization of the permselective properties of excised human skin during iontophoresis, *J. Pharm. Sci.*, **76**, 765–773.

1988, Characterization of the pore transport properties and tissue alteration of excised human skin during iontophoresis, *J. Pharm. Sci.*, **77**, 132–137.

CALDERWOOD, J.H., 1996, Electrode-skin impedance from a dielectric viewpoint, *Physiol. Meas.*, **17**, A131–A139.

CAMEL, E., O'CONNELL, M., SAGE, B., GROSS, M. and MAIBACH, H., 1996, The effect of saline iontophoresis on skin integrity in human volunteers. 1. Methodology and reproducibility, *Fund. Appl. Toxicol.*, **32**, 168–178.

CAMPLEJOHN, R.S., 1996, DNA damage and repair in melanoma and non-melanoma skin cancer, *Cancer Surv.*, **26**, 193–206.

CARROLL, D., TRAMER, M., McQUAY, H., NYE, B. and MOORE, A., 1997, Transcutaneous electrical nerve stimulation in labour pain: A systematic review, *Br. J. Obstet. Gynaecol.*, **104**, 169–175.

CELEBI, N., ERDEN, N., GONUL, B. and KOZ, M., 1994, Effects of epidermal growth factor dosage forms on dermal wound strength in mice, *J. Pharm. Pharmacol.*, **46**, 386–387.

CEVC, G., SCHATZLEIN, A. and BLUME, G., 1995, Transdermal drug carriers: Basic properties, optimization and transfer efficiency in the case of epicutaneously applied peptides, *J. Control. Release*, **36**, 3–16.

CHANG, B.K., GUTHRIE, T.H., HAYAKAWA, K. and GANGAROSA, L.P., 1993, A pilot study of iontophoretic cisplatin chemotherapy of basal and squamous cell carcinomas of the skin, *Arch. Dermatol.*, **129**, 425–427.

CHANTRAINE, A., LUDY, J.P. and BERGER, D., 1986, Is cortisone iontophoresis possible? *Arch. Phys. Med. Rehabil.*, **67**, 38–40.

CHEN, L.H. and CHIEN, Y.W., 1994, Development of a skin permeation cell to simulate clinical study of iontophoretic transdermal delivery, *Drug Develop. Ind. Pharm.*, **20**, 935–945.

1997, Laser scanning confocal microscopic study of transdermal iontophoresis, *Pharm. Res.*, **14**, S307–S308.

CHEN, W. and FRANK, S.G., 1997, Iontophoresis of lidocaine H^+ from poloxamer 407 gels, *Pharm. Res.*, **14**, S309.

CHEN, Y., CHEN, L. and CHIEN, Y.W., 1997a, Insulin denaturation under an electric field, *Pharm. Res.*, **14**, S154.

CHEN, Y., LIN, S. and CHIEN, Y.W., 1997b, Comparative electrochemical stability of human insulin analogues, *Pharm. Res.*, **14**, S153–S154.

CHENG, D.C.H. and VENKATESHWARAN, S., 1991, Iontophoresis device epidermally administering drug – includes rate controlling membrane only permeable to drug when voltage is applied, Pat. WO 9105582. Assignee: Theratech.

CHENG, K., TARJAN, P.P. and MERTZ, P.M., 1996a, Conductivities of pig dermis and subcutaneous fat measured with rectangular pulse electrical current, *Bioelectromagnetics*, **17**, 458–466.

CHENG, K., TARJAN, P.P., THIO, Y.C. and MERTZ, P.M., 1996b, *In vivo* 3-D distributions of electric fields in pig skin with rectangular pulse electrical current stimulation (RPECS), *Bioelectromagnetics*, **17**, 253–262.

CHERNG, J.Y., VAN DE WETERING, P., TALSMA, H., CROMMELIN, D.J.A. and HENNINK, W.E., 1996, Effect of size and serum proteins on transfection efficiency of poly((2-dimethylamino)ethyl methacrylate)-plasmid nanoparticles, *Pharm. Res.*, **13**, 1038–1042.

CHESNOY, S., DOUCET, J., DURAND, D. and COUARRAZE, G., 1996, Effect of iontophoresis in combination with ionic enhancers on the lipid structure of the stratum corneum: An X-ray diffraction study, *Pharm. Res.*, **13**, 1581–1584.

CHIEN, Y.W. and BANGA, A.K., 1989a, Iontophoretic (transdermal) delivery of drugs: Overview of historical development, *J. Pharm. Sci.*, **78**, 353–354.

1989b, Potential developments in systemic delivery of insulin, *Drug Develop. Ind. Pharm.*, **15**, 1601–1634.

1993, Iontotherapeutic devices, reservoir electrode devices therefore, process and unit dose, U.S. Pat. 5250022. Assignee: Rutgers University.

CHIEN, Y.W., SIDDIQUI, O., SUN, Y., SHI, W.M. and LIU, J.C., 1987, Transdermal iontophoretic delivery of therapeutic peptides/proteins: (I) Insulin, in: R.L. JULIANO (ed.), Biological approaches to the controlled delivery of drugs, *Ann. N.Y. Acad. Sci.*, **507**, 32–51.

CHIEN, Y.W., SIDDIQUI, O., SHI, W.M., LELAWONGS, P. and LIU, J.C., 1989, Direct current iontophoretic transdermal delivery of peptide and protein drugs, *J. Pharm. Sci.*, **78**, 376–383.

CHIEN, Y.W., LELAWONGS, P., SIDDIQUI, O., SUN, Y. and SHI, W.M., 1990, Facilitated transdermal delivery of therapeutic peptides and proteins by iontophoretic delivery devices, *J. Control. Release*, **13**, 263–278.

CHOI, H., FLYNN, G.L. and AMIDON, G.L., 1990, Transdermal delivery of bioactive peptides: The effect of n-decylmethyl sulfoxide, pH, and inhibitors on enkephalin metabolism and transport, *Pharm. Res.*, **7**, 1099–1106.

CHU, D.L., CHIOU, H.J. and WANG, D.P., 1994, Characterization of transdermal delivery of nefopam hydrochloride under iontophoresis, *Drug Dev. Ind. Pharm.*, **20**, 2775–2785.

CLANCY, M.J., CORISH, J. and CORRIGAN, O.I. 1994, A comparison of the effects of electrical current and penetration enhancers on the properties of human skin using spectroscopic (FTIR) and calorimetric (DSC) methods, *Int. J. Pharm.*, **105**, 47–56.

CLEMESSY, M., COUARRAZE, G., BEVAN, B. and PUISIEUX, F., 1994, Preservation of skin permeability during *in vitro* iontophoretic experiments, *Int. J. Pharm.*, **101**, 219–226.

1995, Mechanisms involved in iontophoretic transport of angiotensin, *Pharm. Res.*, **12**, 998–1002.

CODERCH, L., OLIVA, M., PONS, L. and PARRA, J.L., 1994, Percutaneous penetration *in vivo* of amino acids, *Int. J. Pharm.*, **111**, 7–14.

COMEAU, M. and BRUMMETT, R., 1978, Anesthesia of the human tympanic membrane by iontophoresis of a local anesthetic, *The Laryngoscope*, **88**, 277–285.

COMEAU, M., BRUMMETT, R. and VERNON, J., 1973, Local anesthesia of the ear by iontophoresis, *Arch. Otolaryngol.*, **98**, 114–120.

COOPER, M.J., 1996, Noninfectious gene transfer and expression systems for cancer gene therapy, *Semin. Oncol.*, **23**, 172–187.

CORDERO, J.A., ALARCON, L., ESCRIBANO, E., OBACH, R. and DOMENECH, J., 1997, A comparative study of the transdermal penetration of a series of nonsteroidal antiinflammatory drugs, *J. Pharm. Sci.*, **86**, 503–508.

CORNWALL, M.W., 1981, Zinc iontophoresis to treat ischemic skin ulcers, *Phys. Ther.*, **61**, 359–360.

COSTELLO, C.T. and JESKE, A.H., 1995, Iontophoresis: Applications in transdermal medication delivery, *Phys. Ther.*, **75**, 554–563.

CRAANE VAN HINSBERG, W.H.M., BAX, L., FLINTERMAN, N.H.M., VERHOEF, J., JUNGINGER, H.E. and BODDE, H.E., 1994, Iontophoresis of a model peptide across human skin *in vitro*: Effects of iontophoresis protocol, pH, and ionic strength on peptide flux and skin impedance, *Pharm. Res.*, **11**, 1296–1300.

CRAANE VAN HINSBERG, I.W.H.M., VERHOEF, J.C., SPIES, F., BOUWSTRA, J.A., GOORIS, G.S., JUNGINGER, H.E. and BODDE, H.E., 1997, Electroperturbation of the human skin barrier *in vitro*. II. Effects on stratum corneum lipid ordering and ultrastructure, *Microsc. Res. Tech.*, **37**, 200–213.

CRAIG, J.A., CUNNINGHAM, M.B., WALSH, D.M., BAXTER, G.D. and ALLEN, J.M., 1996, Lack of effect of transcutaneous electrical nerve stimulation upon experimentally induced delayed onset muscle soreness in humans, *Pain*, **67**, 285–289.

CROOKE, S.T., 1995, Delivery of oligonucleotides and polynucleotides, *J. Drug Targeting*, **3**, 185–190.

CROOKE, S.T., BERNSTEIN, L.S. and BOSWELL, H., 1996, Progress in the development and patenting of antisense drug discovery technology, *Expert. Opin. Ther. Patents.*, **6**, 855–870.

CROSS, S.E. and ROBERTS, M.S., 1995, Importance of dermal blood supply and epidermis on the transdermal iontophoretic delivery of monovalent cations, *J. Pharm. Sci.*, **84**, 584–592.

CRUZ, M.P., EECKHOUDT, S.L., VERBEECK, R.K. and PREAT, V., 1997, Transdermal delivery of flurbiprofen in the rat by iontophoresis and electroporation, *Pharm. Res.*, **14**, S309.

CSILLIK, B., KNYIHAR-CSILLIK, E. and SZUCS, A., 1982, Treatment of chronic pain syndromes with iontophoresis of vinca alkaloids to the skin of patients, *Neurosci. Lett.*, **31**, 87–90.

CUDD, A., ARVINTE, T., DAS, R.E.G., CHINNI, C. and MACINTYRE, A., 1995, Enhanced potency of human calcitonin when fibrillation is avoided, *J. Pharm. Sci.*, **84**, 717–719.

CULLANDER, C., 1992, What are the pathways of iontophoretic current flow through mammalian skin? *Adv. Drug Del. Rev.*, **9**, 119–135.

CULLANDER, C. and GUY, R.H., 1992, (D) Routes of delivery: Case studies (6) Transdermal delivery of peptides and proteins, *Adv. Drug Del. Rev.*, **8**, 291–329.

CULLANDER, C., RAO, G. and GUY, R.H., 1993, Why silver/silver chloride? Criteria for iontophoresis electrodes, in: *Proceedings of the 3rd International Prediction of Percutaneous Penetration Conference: Methods, measurements, modeling*, K.R. BRAIN, V.J. JAMES and K.A. WALTERS (eds). STS Publishing, Cardiff (Wales), pp. 381–390.

D'EMANUELE, A. and STANIFORTH, J.N., 1991, An electrically modulated drug delivery device. I, *Pharm. Res.*, **8**, 913–918.

1992a, An electrically modulated drug delivery device. II. Effect of ionic strength, drug concentration, and temperature, *Pharm. Res.*, **9**, 215–219.

1992b, An electrically modulated drug delivery device. III. Factors affecting drug stability during electrophoresis, *Pharm. Res.*, **9**, 312–315.

DARSOW, U., RING, J., SCHAREIN, E. and BROMM, B., 1996, Correlations between histamine-induced weal, flare and itch, *Arch. Dermatol. Res.*, **288**, 436–441.

DAVIS, H.L., 1996, DNA-based vaccination against hepatitis B virus, *Adv. Drug Deliv. Rev.*, **21**, 33–47.

DAWES, P.T. and FOWLER, P.D. 1995, Treatment of early rheumatoid arthritis: A review of current and future concepts and therapy, *Clin. Exp. Rheumatol.*, **13**, 381–394.

DE BEUKELAAR, F.M.J., MESENS, J.L., VAN REET, G. and VAN RETT, G., 1995, Iontophoretic delivery of anti-migraine drug – particularly dihydro benzopyran-alkylamino alkyl substituted guanidine compounds, Pat. WO 9505815. Assignee: Janssen Pharm.

DELGADOCHARRO, M.B. and GUY, R.H., 1995, Iontophoretic delivery of nafarelin across the skin, *Int. J. Pharm.*, **117**, 165–172.

DELGADOCHARRO, M.B., RODRIGUEZBAYON, A.M. and GUY, R.H., 1995, Iontophoresis of nafarelin: Effects of current density and concentration on electrotransport *in vitro*, *J. Control. Release*, **35**, 35–40.

DENDA, M., KOYAMA, J., NAMBA, R. and HORII, I., 1994, Stratum-corneum lipid morphology and transepidermal water loss in normal skin and surfactant-induced scaly skin, *Arch. Dermatol. Res.*, **286**, 41–46.

DE NUZZIO, J.D. and BERNER, B., 1990, Electrochemical and iontophoretic studies of human skin, *J. Control. Release*, **11**, 105–112.

DE NUZZIO, J., BOERICKE, K., SUTTER, D., MCFARLAND, A., DEY, D., CESARINI, R., MONTY, E., COLVILLE, D., BOCK, R., O'CONNELL, M. and SAGE, B., 1996, Iontophoretic delivery of buprenorphine, *Proc. Int'l Symp. Control. Rel. Bioact. Mater.*, Controlled Release Society, Inc. 23. pp. 285–286.

DE SILVA, O., BASKETTER, D.A., BARRATT, M.D., CORSINI, E., CRONIN, M.T.D., DAS, P.K., DEGWERT, J., ENK, A., GARRIGUE, J.L., HAUSER, C., KIMBER, I., LEPOITTEVIN, J.P., PEGUET, J. and PONEC, M., 1996, Alternative methods for skin sensitisation testing – The report and recommendations of ECVAM Workshop 19, *ATLA*, **24**, 683–705.

DEV, S.B., 1994, Killing cancer cells with a combination of pulsed electric fields and chemotherapeutic agents, *Cancer Watch*, **3**, 12–14.

DEV, S.B. and HOFMANN, G.A., 1994, Electrochemotherapy – a novel method of cancer treatment, *Cancer Treatment Rev.*, **20**, 105–115.

1996, Clinical applications of electroporation, in: *Electrical Manipulation of Cells*, P.T. LYNCH and M.R. DAVEY (eds). Chapman & Hall, London, pp. 185–199.

DICK, I.P. and SCOTT, R.C., 1992, The influence of different strains and age on *in vitro* rat skin permeability to water and mannitol, *Pharm. Res.*, **9**, 884–887.

DINH, S., WOUTERS, S.E. and SCLAFANI, J.R., 1996, Application of thermopiles in iontophoretic drug delivery systems, *Pharm. Res.*, **13**, S360.

DOBSON, R.L., 1987, Treatment of hyperhidrosis, *Arch. Dermatol.*, **123**, 883–884.

DOEGLAS, D., SUURMEIJER, T., KROL, B., SANDERMAN, R., VANLEEUWEN, M. and VANRIJSWIJK, M., 1995, Work disability in early rheumatoid arthritis, *Ann. Rheum. Dis.*, **54**, 455–460.

DONAWA, M.E., 1997, Drug delivery devices: Regulatory and quality requirements for medical devices, Part I, *Pharm. Technol.*, **21**, 110–114.

DONNELLY, J.J., MARTINEZ, D., JANSEN, K.U., ELLIS, R.W., MONTGOMERY, D.L. and LIU, M.A., 1996, Protection against papillomavirus with a polynucleotide vaccine, *J. Infect. Dis.*, **173**, 314–320.

DONNELLY, J.J., ULMER, J.B. and LIU, M.A., 1997, DNA vaccines, *Life Sci.*, **60**, 163–172.

DONNER, B. and ZENZ, M., 1995, Transdermal fentanyl: A new step on the therapeutic ladder, *Anti-Cancer. Drug*, **6**, 39–43.

DREHER, F., WALDE, P., LUISI, P.L. and ELSNER, P., 1996, Human skin irritation studies of a lecithin microemulsion gel and of lecithin liposomes, *Skin Pharmacol.*, **9**, 124–129.

DUCHENE, D. and WOUESSIDJEWE, D., 1990, Pharmaceutical uses of cyclodextrins and derivatives, *Drug Dev. Ind. Pharm.*, **16**, 2487–2499.

DUPLESSIS, J., RAMACHANDRAN, C., WEINER, N. and MULLER, D.G., 1994, The influence of particle size of liposomes on the deposition of drug into skin, *Int. J. Pharm.*, **103**, 277–282.

ECHOLS, D.F., NORRIS, C.H. and TABB, H.G., 1975, Anesthesia of the ear by iontophoresis of lidocaine, *Arch. Otolaryngol.*, **101**, 418–421.

EDWARDS, D.A., PRAUSNITZ, M.R., LANGER, R. and WEAVER, J.C., 1995, Analysis of enhanced transdermal transport by skin electroporation, *J. Control. Release*, **34**, 211–221.

EGBARIA, K., RAMACHANDRAN, C., KITTAYANOND, D. and WEINER, N., 1990, Topical delivery of liposomally encapsulated interferon evaluated by *in vitro* diffusion studies, *Antimicrob. Agents Chemother.*, **34**, 107–110.

EGELRUD, T., LUNDSTROM, A. and SONDELL, B., 1996, Stratum corneum cell cohesion and desquamation in maintenance of the skin barrier, in: *Dermatotoxicology*, F.N. MARZULLI and H.I. MAIBACH (eds). Taylor & Francis, USA, pp. 19–27.

ELGART, M.L. and FUCHS, G., 1987, Tapwater iontophoresis in the treatment of hyperhidrosis, *Int. J. Dermatol.*, **26**, 194–197.

ELIAS, P.M., 1983, Epidermal lipids, barrier function, and desquamation, *J. Invest. Dermatol.*, **80**, 44s–49s.

1988, Structure and function of the stratum corneum permeability barrier, *Drug Dev. Res.*, **13**, 97–105.

EMTESTAM, L. and OLLMAR, S., 1993, Electrical impedance index in human skin: Measurements after occlusion, in five anatomical regions and in mild irritant contact dermatitis, *Contact Dermatitis*, **28**, 104–108.

ERLANGER, G., 1954, Iontophoresis, a scientific and practical tool in ophthalmology, *Ophthalmologica*, **128**, 232–245.

ERNST, A.A., POMERANTZ, J., NICK, T.G., LIMBAUGH, J. and LANDRY, M., 1995, Lidocaine via iontophoresis in laceration repair: A preliminary safety study, *Am. J. Emerg. Med.*, **13**, 17–20.

FABIAN, L. and WILLIAMS, T.J., 1995, Iontophoresis electronic device having a ramped output current, U.S. Pat. 5431625. Assignee: Empi.

FAMAEY, J.P., BROUX, G., CLEPPE, D., DEROULEZ, A., DUCKERTS, F., EVRARD, F., GOETHALS, L., VANHECKE, J., VERBRUGGEN, L. and TYBERGHEIN, J.M., 1982, Ionisation with voltaren: A multi-centre trial, *J. belge Med. Phys. Rehab.*, **5**, 55–60.

FANG, J.Y., HUANG, Y.B., WU, P.C. and TSAI, Y.H., 1996a, Transdermal iontophoresis of sodium nonivamide acetate. I. Consideration of electrical and chemical factors, *Int. J. Pharm.*, **143**, 47–58.

1996b, Transdermal iontophoresis of sodium nonivamide acetate. II. Optimization and evaluation on solutions and gels, *Int. J. Pharm.*, **145**, 175–186.

FANG, J.Y., WU, P.C., HUANG, Y.B. and TSAI, Y.H., 1996c, *In vivo* percutaneous absorption of capsaicin, nonivamide and sodium nonivamide acetate from ointment bases: Pharmacokinetic analysis in rabbits, *Int. J. Pharm.*, **128**, 169–177.

1996d, *In vivo* percutaneous absorption of capsaicin, nonivamide and sodium nonivamide acetate from ointment bases: Skin erythema test and non-invasive surface recovery technique in humans, *Int. J. Pharm.*, **131**, 143–151.

FANG, J.Y., FANG, C.L., HUANG, Y.B. and TSAI, Y.H., 1997, Transdermal iontophoresis of sodium nonivamide acetate. III. Combined effect of pretreatment by penetration enhancers, *Int. J. Pharm.*, **149**, 183–193.

FARTASCH, M., 1996, The nature of the epidermal barrier: Structural aspects, *Adv. Drug Del. Rev.*, **18**, 273–282.

FINKELSTEIN, A. and MAURO, A., 1977, Physical principles and formalisms of electrical excitability, in: *Handbook of Physiology*, Anonymous. Am. Phys. Soc., Bethesda, pp. 161–213.

FIORE, M.C., SMITH, S.S., JORENBY, D.E. and BAKER, T.B., 1994, The effectiveness of the nicotine patch for smoking cessation – A meta-analysis, *JAMA*, **271**, 1940–1947.

FISET, P., COHANE, C., BROWNE, S., BRAND, S.C. and SHAFER, S.L., 1995, Biopharmaceutics of a new transdermal fentanyl device, *Anesthesiology*, **83**, 459–469.

FLOWER, R.J., 1996, Iontophoretic drug delivery device with disposable skin patch and reusable controller – which detects patch compatibility and exhaustion, and the date and time of usage, Pat. WO 9 610 440. Assignee: Becton Dickinson.

FLOWER, R.J. and SAGE, B.H., 1995, Iontophoretic drug delivery system – has two electrodes and associated reservoirs with switch to reverse direction of current flow, Pat. WO 9509032. Assignee: Becton Dickinson.

FLOWER, R.J., MCARTHUR, W.A., STROPKAY, S.E. and TANNER, M.W., 1996, Iontophoretic drug delivery system – has disposable housing with reservoirs and high current power source and removable reusable controller, Pat. WO 9610441. Assignee: Becton Dickinson.

FLYNN, G.L. and STEWART, B., 1988, Percutaneous drug penetration: Choosing candidates for transdermal development, *Drug Dev. Res.*, **13**, 169–185.

FOGT, E.J., NORENBERG, M.S., UNTEREKER, D.F. and COURY, A.J., 1984, Fluid absorbent quantitative test device, U.S. Pat. 4 444 193. Assignee: Medtronic.

1989, Method of preparing a fluid absorbent quantitative test device, U.S. Pat. 4846182. Assignee: Medtronic.

FRANZ, T.J., 1978, The finite dose technique as a valid *in vitro* model for the study of percutaneous absorption in man, *Curr. Probl. Dermatol.*, **7**, 58–68.

FU, J.J. and LORDEN, J.F., 1996, An easily constructed carbon fiber recording and microiontophoresis assembly, *J. Neurosci. Methods*, **68**, 247–251.

FUHRMAN, L.C., MICHNIAK, B.B., BEHL, C.R. and MALICK, A.W., 1997, Effect of novel penetration enhancers on the transdermal delivery of hydrocortisone: An *in vitro* species comparison, *J. Control. Release.*, **45**, 199–206.

FULLER, D.H., MURPHEY CORB, M., CLEMENTS, J., BARNETT, S. and HAYNES, J.R., 1996, Induction of immunodeficiency virus-specific immune responses in rhesus monkeys following gene gun-mediated DNA vaccination, *J. Med. Primatol.*, **25**, 236–241.

FYNAN, E.F., WEBSTER, R.G., FULLER, D.H., HAYNES, J.R., SANTORO, J.C. and ROBINSON, H.L. 1993, DNA Vaccines: Protective immunizations by parenteral, mucosal, and gene-gun inoculations, *Proc. Natl. Acad. Sci.*, **90**, 11478–11482.

GALLO, S.A., OSEROFF, A.R., JOHNSON, P.G. and HUI, S.W., 1997, Characterization of electric-pulse-induced permeabilization of porcine skin using surface electrodes, *Biophys. J.*, **72**, 2805–2811.

GANGA, S., RAMARAO, P. and SINGH, J., 1996, Effect of Azone on the iontophoretic transdermal delivery of metoprolol tartrate through human epidermis *in vitro*, *J. Control. Release*, **42**, 57–64.

GANGAROSA, L.P., 1986, Fluoride iontophoresis for tooth desensitization, *J. Am. Dent. Assoc.*, **112**, 808–810.

GANGAROSA, L.P., PARK, N.H. and HILL, J.M. 1977, Iontophoretic assistance of 5-iodo 2'-deoxyuridine penetration into neonatal mouse skin and effects on DNA synthesis, *Proc. Soc. Exp. Biol. Med.*, **154**, 439–443.

GANGAROSA, L.P., PARK, N.H., FONG, B.C., SCOTT, D.F. and HILL, J.M., 1978, Conductivity of drugs used for iontophoresis, *J. Pharm. Sci.*, **67**, 1439–1443.

GANGAROSA, L.P., MERCHANT, H.W., PARK, N.H. and HILL, J.M., 1979, Iontophoretic application of idoxuridine for recurrent herpes labialis: Report of preliminary clinical findings, *Methods Find. Exp. Clin. Pharmacol.*, **1**, 105–109.

GANGAROSA, L.P., PARK, N., WIGGINS, C.A. and HILL, J.M., 1980, Increased penetration of nonelectrolytes into mouse skin during iontophoretic water transport (iontohydrokinesis), *J. Pharmacol. Exp. Ther.*, **212**, 377–381.

GARAGIOLA, U., DACATRA, U., BRACONARO, F., PORRETTI, E., PISETTI, A. and AZZOLINI, V., 1988, Iontophoretic administration of pirprofen or lysine soluble aspirin in the treatment of rheumatic diseases, *Clinical Ther.*, **10**, 553–558.

GARNSWORTHY, R.K., GULLY, R.L., KENINS, P. and WESTERMAN, R.A., 1988, Transcutaneous electrical stimulation and the sensation of prickle, *J. Neurophys.*, **59**, 1116–1127.

GARRIDO, J., MAFE, S. and PELLICER, J., 1985, Generalization of a finite difference numerical method for the steady state and transient solutions of the Nernst–Planck flux equations, *J. Memb. Sci.*, **24**, 7–14.

GAUTHIER, Y. 1996, Stress and skin: Experimental approaches, *Pathol. Biol.*, **44**, 882–887.

GHOSH, T.K. and BANGA, A.K., 1993a, Methods of enhancement of transdermal drug delivery: Part IIB, Chemical permeation enhancers, *Pharm. Technol.*, **17**(5), 68–76.

1993b, Methods of enhancement of transdermal drug delivery: Part IIA. Chemical permeation enhancers, *Pharm. Technol.*, **17**(4), 62–90.

GIBSON, L.E., 1967, Iontophoretic sweat test for cystic fibrosis: technical details, *Pediatrics*, **39**, 465.

GIBSON, L.E. and COOKE, R.E., 1959, A test for the concentration of electrolytes in sweat in cystic fibrosis of the pancreas utilizing pilocarpine by iontophoresis, *Pediatrics*, **23**, 545–549.

GILBERT, R.A., JAROSZESKI, M.J. and HELLER, R., 1997, Novel electrode designs for electrochemotherapy, *Biochim. Biophys. Acta*, **1334**, 9–14.

GLASS, J.M., STEPHEN, R.L. and JACOBSON, S.C., 1980, The quantity and distribution of radiolabeled dexamethasone delivered to tissue by iontophoresis, *Int. J. Dermatol.*, **19**, 519–525.

GLIKFELD, P., CULLANDER, C., HINZ, R.S. and GUY, R.H., 1988, A new system for *in vitro* studies of iontophoresis, *Pharm. Res.*, **5**, 443–446.

GLIKFELD, P., HINZ, R.S. and GUY, R.H., 1989, Noninvasive sampling of biological fluids by iontophoresis, *Pharm. Res.*, **6**, 988–990.

GLIKFELD, P., CULLANDER, C., HINZ, R.S. and GUY, R.H., 1994, Device for iontophoretic non-invasive sampling or delivery of substances, U.S. Pat. 5279543. Assignee: Univ. of California.

GODBEY, K.J., 1997, Development of a novel transdermal drug delivery backing film with a low moisture vapor transmission rate, *Pharm. Technol.*, **21**, 98–107.

GOLDMAN, R. and POLLACK, S., 1996, Electric fields and proliferation in a chronic wound model, *Bioelectromagnetics*, **17**, 450–457.

GREEN, P., 1996a, Iontophoretic transdermal drug delivery: A new commercially feasible technology, Hotel International, Basel, Switzerland, October 17–18 (Organized by A.K. BANGA and P. GREEN through Technomic Publishing).

1996b, Iontophoretic delivery of peptide drugs, *J. Control. Release*, **41**, 33–48.

GREEN, P., SHROOT, B., BERNERD, F., PILGRIM, W.R. and GUY, R.H., 1992, *In vitro* and *in vivo* iontophoresis of a tripeptide across nude rat skin, *J. Control. Release*, **20**, 209–218.

GREEN, P.G., HADGRAFT, J. and RIDOUT, G., 1989, Enhanced *in vitro* skin permeation of cationic drugs, *Pharm. Res.*, **6**, 628–632.

GREEN, P.G., HINZ, R.S., CULLANDER, C., YAMANE, G. and GUY, R.H., 1991a, Iontophoretic delivery of amino acids and amino acid derivatives across the skin *in vitro*, *Pharm. Res.*, **8**, 1113–1120.

GREEN, P.G., HINZ, R.S., KIM, A., SZOKA, F.C. and GUY, R.H., 1991b, Iontophoretic delivery of a series of tripeptides across the skin *in vitro*, *Pharm. Res.*, **8**, 1121–1127.

GREEN, P.G., FLANAGAN, M., SHROOT, B. and GUY, R.H., 1993, Iontophoretic drug delivery, in: *Pharmaceutical Skin Penetration Enhancement*, K.A. WALTERS and J. HADGRAFT (eds). Marcel Dekker Inc., New York, pp. 311–333.

GREENBAUM, S.S. and BERNSTEIN, E.F., 1994, Comparison of iontophoresis of lidocaine with a eutectic mixture of lidocaine and prilocaine (EMLA) for topically administered local anesthesia, *J. Dermatol. Surg. Oncol.*, **20**, 579–583.

GREGORIADIS, G., SAFFIE, R. and DE SOUZA, J.B., 1997, Liposome-mediated DNA vaccination, *FEBS Lett.*, **402**, 107–110.

GREMINGER, R.F., ELLIOTT, R.A. and RAPPERPORT, A., 1980, Antibiotic iontophoresis for the management of burned ear chondritis, *Plastic Reconst. Surg.*, **66**, 356–359.

GRICE, K., SATTAR, H. and BAKER, H., 1972, Treatment of idiopathic hyperhidrosis with iontophoresis of tap water and poldine methosulfate, *Br. J. Dermatol.*, **86**, 72–78.

GRIMNES, S., 1984, Pathways of ionic flow through human skin *in vivo*, *Acta. Derm. Venereol. (Stockh.)*, **64**, 93–98.

GRONING, R., 1987, Electrophoretically controlled dermal or transdermal application systems with electronic indicators, *Int. J. Pharm.*, **36**, 37–40.

GRONINGSSON, K., LINDGREN, J.E., LUNDBERG, E., SANDBERG, R. and WAHLEN, A., 1985, Lidocaine base and hydrochloride, *Anal. Prof. Drug Subst.*, **14**, 207–243.

GUPTA, S.K., KUMAR, S., BOLTON, S., BEHL, C.R. and MALICK, A.W., 1994a, Effect of chemical enhancers and conducting gels on iontophoretic transdermal delivery of cromolyn sodium, *J. Control. Release*, **31**, 229–236.

1994b, Optimization of iontophoretic transdermal delivery of a peptide and a non-peptide drug, *J. Control. Release*, **30**, 253–261.

GUY, R.H., 1995, A sweeter life for diabetics? *Nature Med.*, **1**, 1132–1133.

1996, Current status and future prospects of transdermal drug delivery, *Pharm. Res.*, **13**, 1765–1769.

GUZEK, D.B., KENNEDY, A.H., MCNEILL, S.C., WAKSHULL, E. and POTTS, R.O., 1989, Transdermal drug transport and metabolism. I. Comparison of *in vitro* and *in vivo* results, *Pharm. Res.*, **6**, 33–39.

GWYNNE, P., 1997, Companies developing more uses for iontophoresis, *The Scientist*, **11**, 1.

GYORY, J.R. and PERRY, J.R., 1994, Iontophoretic delivery device and method of hydrating same, U.S. Pat. 5310404. Assignee: Alza Corp.

GYORY, J.R., MOODIE, L.C., PHIPPS, J.B. and THEEUWES, F., 1995, Electrically powered iontophoretic device for enhanced drug delivery – has reservoir for drug, pref fentanyl, in electrode and ion exchange material with mobile and immobile ion species in same or other electrode, Pat. WO 9527529. Assignee: Alza Corp.

HAAK, R.P., MYERS, R.M. and PLUE, R.W., 1997, Iontophoretic drug delivery device – has electrical pathway formed on one side of a flexible, non-conducting substrate, Pat. EP 751801. Assignee: Alza Corp.

HAGA, M., AKATANI, M., KIKUCHI, J., UENO, Y. and HAYASHI, M., 1997, Transdermal iontophoretic delivery of insulin using a photoetched microdevice, *J. Control. Release*, **43**, 139–149.

HAGER, D.F., LAUBACH, M.J., SHARKEY, J.W. and SIVERLY, J.R., 1993, In vitro iontophoretic delivery of CQA 206-291 – Influence of ethanol, *J. Control. Release.*, **23**, 175–182.

HAGER, D.F., MANCUSO, F.A., NAZARENO, J.P., SHARKEY, J.W. and SIVERLY, J.R., 1994, Evaluation of a cultured skin equivalent as a model membrane for iontophoretic transport, *J. Control. Release*, **30**, 117–123.

HANSCH, R., KOPREK, T., HEYDEMANN, H., MENDEL, R.R. and SCHULZE, J., 1996, Electroporation-mediated transient gene expression in isolated scutella of *Hordeum vulgare*, *Physiol. Plant*, **98**, 20–27.

HAO, J., LI, D., LI, S. and ZHENG, J., 1996, The effects of some penetration enhancers on the transdermal iontophoretic delivery of insulin in vitro, *J. Chin. Pharm. Sci.*, **5**, 88–92.

HARDEn, J.L. and VIOVY, J.L., 1996, Numerical studies of pulsed iontophoresis through model membranes, *J. Control. Release*, **38**, 129–139.

HARDING, J.W. and FELIX, D., 1987, Quantification of angiotensin iontophoresis, *J. Neurosc. Methods*, **19**, 209–215.

HARRIS, P.R., 1982, Iontophoresis: Clinical research in musculoskeletal inflammatory conditions, *J. Orthop. Sports Phys. Ther.*, **4**, 109–112.

HASSON, S.H., HENDERSON, G.H., DANIELS, J.C. and SCHIEB, D.A., 1991, Exercise training and dexamethasone iontophoresis in rheumatoid arthritis, *Physiotherapy Canada*, **43**, 11–14.

HASSON, S.M., WIBLE, C.L., REICH, M., BARNES, W.S. and WILLIAMS, J.H., 1992, Dexamethasone iontophoresis: Effect on delayed muscle soreness and muscle function, *Can. J. Spt. Sci.*, **17**, 8–13.

HATANAKA, T., KAMON, T., UOZUMI, C., MORIGAKI, S., AIBA, T., KATAYAMA, K. and KOIZUMI, T., 1996, Influence of pH on skin permeation of amino acids, *J. Pharm. Pharmacol.*, **48**, 675–679.

HAYNES, J.L., 1992, Method for obtaining blood using iontophoresis, U.S. Pat. 5131403. Assignee: Becton Dickinson.

1994, Failsafe iontophoresis drug delivery system, U.S. Pat. 5306235. Assignee: Becton Dickinson.

HAYNES, J.R., MCCABE, D.E., SWAIN, W.F., WIDERA, G. and FULLER, J.T., 1996, Particle-mediated nucleic acid immunization, *J. Biotechnol.*, **44**, 37–42.

HEARD, C.M. and BRAIN, K.R., 1995, Does solute stereochemistry influence percutaneous penetration? *Chirality*, **7**, 305–309.

HEIT, M.C., WILLIAMS, P.L., JAYES, F.L., CHANG, S.K. and RIVIERE, J.E., 1993, Transdermal iontophoretic peptide delivery – In vitro and in vivo studies with luteinizing hormone releasing hormone, *J. Pharm. Sci.*, **82**, 240–243.

HEIT, M.C., MCFARLAND, A., BOCK, R. and RIVIERE, J.E., 1994a, Isoelectric focusing and capillary zone electrophoretic studies using luteinizing hormone releasing hormone and its analog, *J. Pharm. Sci.*, **83**, 654–656.

HEIT, M.C., MONTEIRORIVIERE, N.A., JAYES, F.L. and RIVIERE, J.E., 1994b, Transdermal iontophoretic delivery of luteinizing hormone releasing hormone (LHRH): Effect of repeated administration, *Pharm. Res.*, **11**, 1000–1003.

HELLER, R., JAROSZESKI, M.J., GLASS, L.F., MESSINA, J.L., RAPAPORT, D.P., DE CONTI, R.C., FENSKE, N.A., GILBERT, R.A., MIR, L.M. and REINTGEN, D.S., 1996, Phase I/II trial for the treatment of cutaneous and subcutaneous tumors using electrochemotherapy, *Cancer*, **77**, 964–971.

HELLER, R., JAROSZESKI, M., PERROTT, R., MESSINA, J. and GILBERT, R., 1997, Effective treatment of B16 melanoma by direct delivery of bleomycin using electrochemotherapy, *Melanoma Res.*, **7**, 10–18.

Bibliography

HENLEY, E.J., 1991, Transcutaneous drug delivery: Iontophoresis, phonophoresis, *Crit. Rev. Ther. Drug Carr. Syst.*, **2**, 139–151.

HENLEY-COHN, J. and HAUSFELD, J.N., 1984, Iontophoretic treatment of oral herpes, *Laryngoscope*, **94**, 118–121.

HERRMANN, J.E., CHEN, S.C., FYNAN, E.F., SANTORO, J.C., GREENBERG, H.B., WANG, S.X. and ROBINSON, H.L., 1996, Protection against rotavirus infections by DNA vaccination, *J. Infect. Dis.*, **174**, S93–S97.

HIGO, N., IGA, K., MATSUMOTO, Y. and YANAI, S., 1996, Iontophoresis interface useful in delivery of peptide drugs with high bioavailability – comprises hydrophilised fluro-resin membrane with low protein adsorptivity, Pat. EP 748636. Assignee: Hisamitsu; Takeda.

HIGOUNENC, I., DEMARCHEZ, M., REGNIER, M., SCHMIDT, R., PONEC, M. and SHROOT, B., 1994, Improvement of epidermal differentiation and barrier function in reconstructed human skin after grafting onto athymic nude mice, *Arch. Dermatol. Res.*, **286**, 107–114.

HILL, A.C., BAKER, G.F. and JANSEN, G.T., 1981, Mechanism of action of iontophoresis in the treatment of palmar hyperhidrosis, *Cutis*, **28**, 69–72.

HILL, J.M., GANGAROSA, L.P. and PARK, N.H., 1977, Iontophoretic application of antiviral chemotherapeutic agents, *Ann. N.Y. Acad. Sci.*, **284**, 604.

HILL, J.M., O'CALLAGHAN, R.J. and HOBDEN, J.A., 1993, Ocular iontophoresis, in: *Ophthalmic Drug Selivery Systems*, A.K. MITRA (ed). Marcel Dekker, Inc., New York, pp. 331–354.

HINSBERG, W.H.M.C., VERHOEF, J.C., BAX, L.J., JUNGINGER, H.E. and BODDE, H.E., 1995, Role of appendages in skin resistance and iontophoretic peptide flux: Human versus snake skin, *Pharm. Res.*, **12**, 1506–1512.

HIRVONEN, J. and GUY, R.H., 1997a, Attenuation of electro-osmotic flow during transdermal iontophoresis. I. Effect of poly-L-lysines, *Proc. Int'l Symp. Control. Rel. Bioact. Mater.*, Controlled Release Society, Inc., 24, pp. 689–690.

1997b, Attenuation of electro-osmotic flow during transdermal iontophoresis. II. Effect of beta-blocking agents, *Proc. Int'l Symp. Control. Rel. Bioact. Mater.*, Controlled Release Society, Inc., 24, pp. 691–692.

1997c, Iontophoretic delivery across the skin: Electroosmosis and its modulation by drug substances, *Pharm. Res.*, **14**, 1258–1263.

HIRVONEN, J., KONTTURI, K., MURTOMAKI, L., PARONEN, P. and URTTI, A., 1993, Transdermal iontophoresis of sotalol and salicylate – The effect of skin charge and penetration enhancers, *J. Control. Release*, **26**, 109–117.

HIRVONEN, J., HUEBER, F. and GUY, R.H., 1995, Current profile regulates iontophoretic delivery of amino acids across the skin, *J. Control. Release*, **37**, 239–249.

HIRVONEN, J., KALIA, Y.N. and GUY, R.H., 1996, Transdermal delivery of peptides by iontophoresis, *Nat. Biotechnol.*, **14**, 1710–1713.

HISOIRE, G. and BUCKS, D., 1997, An unexpected finding in percutaneous absorption observed between haired and hairless guinea pig skin, *J. Pharm. Sci.*, **86**, 398–400.

HO, H. and CHIEN, Y.W., 1993, Kinetic evaluation of transdermal nicotine delivery systems, *Drug Develop. Ind. Pharm.*, **19**, 295–313.

HOFLAND, H.E.J., BOUWSTRA, J.A., BODDE, H.E., SPIES, F. and JUNGINGER, H.E., 1995, Interactions between liposomes and human stratum corneum *in vitro*: freeze fracture electron microscopical visualization and small angle X-ray scattering studies, *Br. J. Dermatol.*, **132**, 853–856.

HOFMANN, G.A., 1995a, Instrumentation, in: *Electroporation Protocols for Microorganisms*, J.A. NICKOLOFF, (ed). Humana Press, Totowa, pp. 27–45.

1995b, Transdermal drug delivery by electroincorporation of vesicles, U.S. Pat. 5464386. Assignee: Genetronics.

1995c, Electroporation system with voltage control feedback for clinical applications, U.S. Pat. 5439440. Assignee: Genetronics.

HOFMANN, G.A., RUSTRUM, W.V. and SUDER, K.S., 1995, Electro-incorporation of microcarriers as a method for the transdermal delivery of large molecules, *Bioelectrochem. Bioenerg.*, **38**, 209–222.

HOLLADAY, L.A., TREAT-CLEMONS, L.G. and BASSETT, P.M., 1996, Composition and method for enhancing electrotransport agent delivery, Pat. WO 96/15826. Assignee: Alza Corp.

HOLLINGER, M.A., 1996, Toxicological aspects of topical silver pharmaceuticals, *Crit. Rev. Toxicol.*, **26**, 255–260.

HOLZLE, E. and RUZICKA, T., 1986, Treatment of hyperhidrosis by a battery operated iontophoretic device, *Dermatologica*, **172**, 41–47.

HOOGSTRAATE, A.J., VERHOEF, J., BRUSSEE, J., IJZERMAN, A.P., SPIES, F. and BODDE, H.E., 1991, Kinetics, ultrastructural aspects and molecular modelling of transdermal peptide flux enhancement by N-alkylazacycloheptanones, *Int. J. Pharm.*, **76**, 37–47.

HOOGSTRAATE, A.J., SRINIVASAN, V., SIMS, S.M. and HIGUCHI, W.I., 1994, Iontophoretic enhancement of peptides: Behaviour of leuprolide versus model permeants, *J. Control. Release*, **31**, 41–47.

HOPSU-HAVU, V.K., FRAKI, J.E. and JARVINEN, M., 1977, Proteolytic enzymes in the skin, in: *Proteinases in Mammalian Cells and Tissues*, Anonymous. North-Holland Biomedical Press, pp. 547–581.

HORI, M., MAIBACH, H.I. and GUY, R.H., 1992, Enhancement of propranolol hydrochloride and diazepam skin absorption *in vitro*. II. Drug, vehicle, and enhancer penetration kinetics, *J. Pharm. Sci.*, **81**, 330–333.

HOWARD, J.P., DRAKE, T.R. and KELLOGG, D.L., 1995, Effects of alternating current iontophoresis on drug delivery, *Arch. Phys. Med. Rehab.*, **76**, 463–466.

HOWES, D., GUY, R., HADGRAFT, J., HEYLINGS, J., HOECK, U., KEMPER, F., MAIBACH, H., MARTY, J.P., MERK, H., PARRA, J., REKKAS, D., RONDELLI, I., SCHAEFER, H., TAUBER, U. and VERBIESE, N., 1996, Methods for assessing percutaneous absorption – The report and recommendations of ECVAM Workshop 13, *ATLA. Altern. Lab. Anim.*, **24**, 81–106.

HUANG, Y.Y. and WU, S.M., 1996a, Transdermal iontophoretic delivery of thyrotropin-releasing hormone across excised rabbit pinna skin, *Drug Develop. Ind. Pharm.*, **22**, 1075–1081.

1996b, Stability of peptides during iontophoretic transdermal delivery, *Int. J. Pharm.*, **131**, 19–23.

HUANG, Y.Y., WU, S.M., WANG, C.Y. and JIANG, T.S., 1995, A strategy to optimize the operation conditions in iontophoretic transdermal delivery of pilocarpine, *Drug Develop. Ind. Pharm.*, **21**, 1631–1648.

HUANG, Y.Y., WU, S.M. and WANG, C.Y., 1996, Response surface method: A novel strategy to optimize iontophoretic transdermal delivery of thyrotropin-releasing hormone, *Pharm. Res.*, **13**, 547–552.

HUDSON, A.J., LEE, W., PORTER, J., AKHTAR, J., DUNCAN, R. and AKHTAR, S., 1996, Stability of antisense oligonucleotides during incubation with a mixture of isolated lysosomal enzymes, *Int. J. Pharm.*, **133**, 257–263.

HUFF, B.V., LIVERSIDGE, G.G. and MCINTIRE, G.L., 1995, The electrophoretic mobility of tripeptides as a function of pH and ionic strength: Comparison with iontophoretic flux data, *Pharm. Res.*, **12**, 751–755.

ILLEL, B., 1997, Formulation for transfollicular drug administration: Some recent advances, *Crit. Rev. Ther. Drug Carr. Syst.*, **14**, 207–219.

INADA, H., GHANEM, A.H. and HIGUCHI, W.I., 1994, Studies on the effects of applied voltage and duration on human epidermal membrane alteration/recovery and the resultant effects upon iontophoresis, *Pharm. Res.*, **11**, 687–697.

INAMORI, T., GHANEM, A.H., HIGUCHI, W.I. and SRINIVASAN, V., 1994, Macromolecule transport in and effective pore size of ethanol pretreated human epidermal membrane, *Int. J. Pharm.*, **105**, 113–123.

INOUE, N., KOBAYASHI, D., KIMURA, M., TOYAMA, M., SUGAWARA, I., ITOYAMA, S., OGIHARA, M., SUGIBAYASHI, K. and MORIMOTO, Y., 1996, Fundamental investigation of a novel drug delivery system, a transdermal delivery system with jet injection, *Int. J. Pharm.*, **137**, 75–84.

IRWIN, W.J., SANDERSON, F.D. and PO, A.L.W., 1990, Percutaneous absorption of ibuprofen and naproxen: Effect of amide enhancers on transport through rat skin, *Int. J. Pharm.*, **66**, 243–252.

JADOUL, A., HANCHARD, C., THYSMAN, S. and PREAT, V., 1995, Quantification and localization of fentanyl and TRH delivered by iontophoresis in the skin, *Int. J. Pharm.*, **120**, 221–228.

JADOUL, A., MESENS, J., CAERS, W., BEUKELAAR, F., CRABBE, R. and PREAT, V., 1996, Transdermal permeation of alniditan by iontophoresis: *In vitro* optimization and human pharmacokinetic data, *Pharm. Res.*, **13**, 1348–1353.

JADOUL, A., REGNIER, V. and PREAT, V., 1997, Influence of ethanol and propylene glycol addition on the transdermal delivery by iontophoresis and electroporation, *Pharm. Res.*, **14**, S308–S309.

JAMES, M.P., GRAHAM, R.M. and ENGLISH, J., 1986, Percutaneous iontophoresis of prednisolone – a pharmacokinetic study, *Clin. Exp. Dermatol.*, **11**, 54–61.

JAROSZESKI, M.J., GILBERT, R., PERROTT, R. and HELLER, R., 1996, Enhanced effects of multiple treatment electrochemotherapy, *Melanoma Res.*, **6**, 427–433.

JAROSZESKI, M.J., GILBERT, R.A. and HELLER, R., 1997, *In vivo* antitumor effects of electrochemotherapy in a hepatoma model, *Biochim. Biophys. Acta*, **1334**, 15–18.

JASS, J., COSTERTON, J.W. and LAPPIN-SCOTT, H.M., 1995, The effect of electrical currents and tobramycin on *Pseudomonas aeruginosa* biofilms, *J. Ind. Microb.*, **15**, 234–242.

JAW, F.S., WANG, C.Y. and HUANG, Y.Y., 1995, Portable current stimulator for transdermal iontophoretic drug delivery, *Med. Eng. Phys.*, **17**, 385–386.

JENKINSON, D.M. and WALTON, G.S., 1974, The potential use of iontophoresis in the treatment of skin disorders, *Vet. Rec.*, **94**, 8.

JENKINSON, D.M., HUTCHISON, G., JACKSON, D. and McQUEEN, L., 1986, Route of passage of cypermethrin across the surface of sheep skin, *Res. Vet. Sci.*, **41**, 237–241.

JOHNSON, M.T.V. and LEE, N.H., 1990, pH buffered electrodes for medical iontophoresis, U.S. Pat. 4973303. Assignee: Empi.

JONES, B., LYNCH, P.T., POWER, J.B. and DAVEY, M.R., 1996, Electrofusion and electroporation equipment, in: *Electrical Manipulation of Cells*. P.T. LYNCH and M.K. DAVEY, (eds). Chapman & Hall, pp. 1–14.

JONG, Y.S., JACOB, J.S., YIP, K.P., GARDNER, G., SEITELMAN, E., WHITNEY, M., MONTGOMERY, S. and MATHIOWITZ, E., 1997, Controlled release of plasmid DNA, *J. Control. Release*, **47**, 123–134.

KAHN, J., 1977, Acetic acid iontophoresis for calcium deposits, *Phys. Ther.*, **57**, 658–659.

1982, A case report: Lithium iontophoresis for gouty arthritis, *J. Orthop. Sports Phys. Ther.*, **4**, 113–114.

KALIA, Y.N. and GUY, R.H., 1995, The electrical characteristics of human skin *in vivo*, *Pharm. Res.*, **12**, 1605–1613.

1997, Interaction between penetration enhancers and iontophoresis: Effect on human skin impedance *in vivo*, *J. Control. Release*, **44**, 33–42.

KALIA, Y.N., NONATO, L.B. and GUY, R.H., 1996, The effect of iontophoresis on skin barrier integrity: Non-invasive evaluation by impedance spectroscopy and transepidermal water loss, *Pharm. Res.*, **13**, 957–960.

KAMATH, S.S. and GANGAROSA, L.P., 1995, Electrophoretic evaluation of the mobility of drugs suitable for iontophoresis, *Meth. Find. Exp. Clin. Pharmacol.*, **17**, 227–232.

KAMBE, M., ARITA, D., KIKUCHI, H., FUNATO, T., TEZUKA, F., GAMO, M., MURAKAWA, Y. and KANAMARU, R., 1996, Enhancing the effect of anticancer drugs against the colorectal cancer cell line with electroporation, *Tohoku. J. Exp. Med.*, **180**, 161–171.

KANJILAL, S., STROM, S.S., CLAYMAN, G.L., WEBER, R.S., ELNAGGAR, A.K., KAPUR, V., CUMMINGS, K.K., HILL, L.A., SPITZ, M.R., KRIPKE, M.L. and ANANTHASWAMY, H.N. 1995, p53 mutations in nonmelanoma skin cancer of the head and neck: Molecular evidence for field cancerization, *Cancer Res.*, **55**, 3604–3609.

KARAMI, K., SJOBERG, H. and BERONIUS, P., 1997, Ionization conditions for iontophoretic drug delivery. Electrical conductance and aggregation of lidocaine hydrochloride in 1-octanol at 25 degrees C, *Int. J. Pharm.*, **154**, 79–87.

KARI, B., 1986, Control of blood glucose levels in alloxan-diabetic rabbits by iontophoresis of insulin, *Diabetes*, **35**, 217–221.

KASSAN, D.G., LYNCH, A.M. and STILLER, M.J., 1996, Physical enhancement of dermatologic drug delivery: Iontophoresis and phonophoresis, *J. Am. Acad. Dermatol.*, **34**, 657–666.

KASTING, G.B. 1992, Theoretical models for iontophoretic delivery, *Adv. Drug Del. Rev.*, **9**, 177–199.

KASTING, G.B. and BOWMAN, L.A., 1990a, DC electrical properties of frozen, excised human skin, *Pharm. Res.*, **7**, 134–143.

1990b, Electrical analysis of fresh, excised human skin: A comparison with frozen skin, *Pharm. Res.*, **7**, 1141–1146.

KASTING, G.B. and KEISTER, J.C., 1989, Application of electrodiffusion theory for a homogeneous membrane to iontoporetic transport through skin, *J. Control. Release*, **8**, 195–210.

KASTING, G.B., MERRITT, E.W. and KEISTER, J.C., 1988, An *in vitro* method for studying the iontophoretic enhancement of drug transport through skin, *J. Memb. Sci.*, **35**, 137–159.

KEISTER, J.C. and KASTING, G.B., 1986, Ionic mass transport through a homogeneous membrane in the presence of a uniform electric field, *J. Memb. Sci.*, **29**, 155–167.

KENNARD, C.D. and WHITAKER, D.C., 1992, Iontophoresis of lidocaine for anesthesia during pulsed dye laser treatment of port-wine stains, *J. Dermatol. Surg. Oncol.*, **18**, 287–294.

KIM, A., GREEN, P.G., RAO, G. and GUY, R.H., 1993, Convective solvent flow across the skin during iontophoresis, *Pharm. Res.*, **10**, 1315–1320.

KIMOTO, H. and TAKETO, A., 1996, Studies on electrotransfer of DNA into *Escherichia coli*: Effect of molecular form of DNA, *Biochim. Biophys. Acta*, **1307**, 325–330.

KIRCHNER, L.A., MOODY, R.P., DOYLE, E., BOSE, R., JEFFERY, J. and CHU, I., 1997, The prediction of skin permeability by using physicochemical data, *ATLA*, **25**, 359–370.

KIRJAVAINEN, M., URTTI, A., JAASKELAINEN, I., SUHONEN, T.M., PARONEN, P., VALJAKKA KOSKELA, R., KIESVAARA, J. and MONKKONEN, J., 1996, Interaction of liposomes with human skin *in vitro* – The influence of lipid composition and structure, *Biochim. Biophys. Acta*, **1304**, 179–189.

KITAGAWA, S., YOKOCHI, N. and MUROOKA, N., 1995, pH-dependence of phase transition of the lipid bilayer of liposomes of stratum corneum lipids, *Int. J. Pharm.*, **126**, 49–56X.

KIYOHARA, Y., KOMADA, F., IWAKAWA, S., FUWA, T. and OKUMURA, K., 1993, Systemic effects of epidermal growth factor (EGF) ointment containing protease inhibitor or gelatin in rats with burns or open wounds, *Biol. Pharm. Bull.*, **16**, 73–76.

KNOBLAUCH, P. and MOLL, F., 1993, *In vitro* pulsatile and continuous transdermal delivery of buserelin by iontophoresis, *J. Control. Release*, **26**, 203–212.

KOBAYASHI, D., MATSUZAWA, T., SUGIBAYASHI, K., MORIMOTO, Y., KOBAYASHI, M. and KIMURA, M., 1993, Feasibility of use of several cardiovascular agents in transdermal therapeutic systems with l-menthol-ethanol system on hairless rat and human skin, *Biol. Pharm. Bull.*, **16**, 254–258.

KOBAYASHI, I., HOSAKA, K., UENO, T., MARUO, H., KAMIYAMA, M., KONNO, C. and GEMBA, M., 1997, Relationship between amount of beta-blockers permeating through the stratum corneum and skin irritation after application of beta-blocker adhesive patches to guinea pig skin, *Biol. Pharm. Bull.*, **20**, 421–427.

KOCHAK, G.M., SUN, J.X., CHOI, R.L. and PIRAINO, A.J., 1992, Pharmacokinetic disposition of multiple-dose transdermal nicotine in healthy adult smokers, *Pharm. Res.*, **9**, 1451–1455.

KONTTURI, K. and MURTOMAKI, L., 1996, Mechanistic model for transdermal transport including iontophoresis, *J. Control. Release*, **41**, 177–185.

KONTTURI, K., MURTOMAKI, L., HIRVONEN, J., PARONEN, P. and URTTI, A., 1993, Electrochemical characterization of human skin by impedance spectroscopy: The effect of penetration enhancers, *Pharm. Res.*, **10**, 381–385.

KORTE, W., DE STOUTZ, N. and MORANT, R., 1996, Day-to-day titration to initiate transdermal fentanyl in patients with cancer pain: Short- and long-term experiences in a prospective study of 39 patients, *J. Pain Symptom. Mgmt.*, **11**, 139–146.

KOST, J., PLIQUETT, U., MITRAGOTRI, S., YAMAMOTO, A., LANGER, R. and WEAVER, J., 1996, Synergistic effect of electric field and ultrasound on transdermal transport, *Pharm. Res.*, **13**, 633–638.

Bibliography

KOU, J.H., ROY, S.D., DU, J. and FUJIKI, J., 1993, Effect of receiver fluid pH on *in vitro* skin flux of weakly ionizable drugs, *Pharm. Res.*, **10**, 986–990.

KRISHNA, R. and PANDIT, J.K., 1996, Carboxymethylcellulose-sodium based transdermal drug delivery system for propranolol, *J. Pharm. Pharmacol.*, **48**, 367–370.

KULKARNI, S.B., BANGA, A.K. and BETAGERI, G.V., 1996, Transdermal iontophoretic delivery of colchicine encapsulated in liposomes, *Drug Del.*, **3**, 245–250.

KUMAR, S., CHAR, H., PATEL, S., PIEMONTESE, D., IQBAL, K., MALICK, A.W., NEUGROSCHEL, E. and BEHL, C.R., 1992, Effect of iontophoresis on *in vitro* skin permeation of an analogue of growth hormone releasing factor in the hairless guinea pig model, *J. Pharm. Sci.*, **81**, 635–639.

KUSHLA, G.P. and ZATZ, J.L., 1991, Influence of pH on lidocaine penetration through human and hairless mouse skin *in vitro*, *Int. J. Pharm.*, **71**, 167–173.

KUZMIN, P.I., DARMOSTUK, A.S., CHIZMADZHEV, Y.A., WHITE, H.S. and POTTS, R.O., 1996, A mechanism of skin appendage macropores electroactivation during iontophoresis, *Membr. Cell Biol.*, **10**, 699–706.

KWON, B.S., HILL, J.M., WIGGINS, C., TUGGLE, C. and GANGAROSA, L.P., 1979, Iontophoretic application of adenosine arabinoside monophosphate for the treatment of herpes simplex virus type 2 infections in hairless mice, *J. Infect. Dis.*, **140**, 1014.

LABHASETWAR, V. and LEVY, R.J., 1995, Novel delivery of antiarrhythmic agents, *Clin. Pharmacokinet.*, **29**, 1–5.

LABHASETWAR, V., UNDERWOOD, T., SCHWENDEMAN, S.P. and LEVY, R.J., 1995, Iontophoresis for modulation of cardiac drug delivery in dogs, *Proc. Natl. Acad. Sci. USA*, **92**, 2612–2616.

LA FOREST, N.T. and CONFRANCESCO, C., 1978, Antibiotic iontophoresis in the treatment of ear chronditis, *Phys. Ther.*, **58**, 32–34.

LANE, J.D., WESTMAN, E.C., RIPKA, G.V., WU, J.L., CHIANG, C.C. and ROSE, J.E., 1993, Pharmacokinetics of a transdermal nicotine patch compared to nicotine gum, *Drug Dev. Ind. Pharm.*, **19**, 1999–2010.

LANGKJAER, L., BRANGE, J., GRODSKY, G.M. and GUY, R.H., 1994, Transdermal delivery of monomeric insulin analogues by iontophoresis, *Proc. Int'l Symp. Control. Rel. Bioact. Mater.*, Controlled Release Society, Inc., 21, pp. 172–173.

LASHMAR, U.T. and MANGER, J., 1994, Investigation into the potential for iontophoresis facilitated transdermal delivery of acyclovir, *Int. J. Pharm.*, **111**, 73–82.

LASIC, D.D. and TEMPLETON, N.S., 1996, Liposomes in gene therapy, *Adv. Drug Del. Rev.*, **20**, 221–266.

LATTIN, G.A. and SPEVAK, R.P., 1984, Non-invasive diagnosis method, U.S. Pat. 4457748. Assignee: Medtronic.

LATTIN, G.A., PADMANABHAN, R.V. and PHIPPS, J.B., 1991, Electronic control of iontophoretic drug delivery, *Ann. N.Y. Acad. Sci.*, **618**, 450–464.

LAU, D.T.W., SHARKEY, J.W., PETRYK, L., MANCUSO, F.A., YU, Z.L. and TSE, F.L.S., 1994, Effect of current magnitude and drug concentration on iontophoretic delivery of octreotide acetate (Sandostatin(R)) in the rabbit, *Pharm. Res.*, **11**, 1742–1746.

LAUER, A.C., LIEB, L.M., RAMACHANDRAN, C., FLYNN, G.L. and WEINER, N.D., 1995, Transfollicular drug delivery, *Pharm. Res.*, **12**, 179–186.

LAYMAN, P.R., ARGYRAS, E. and GLYNN, C.J., 1986, Iontophoresis of vincristine versus saline in post-herpetic neuralgia. A controlled trial, *Pain*, **25**, 165–170.

LEDGER, P.W., 1992, Skin biological issues in electrically enhanced transdermal delivery, *Adv. Drug Del. Rev.*, **9**, 289–307.

LEE, A.J., KING, J.R. and BARRETT, D.A., 1997, Percutaneous absorption: A multiple pathway model, *J. Control. Release.*, **45**, 141–151.

LEE, Y.M., SONG, K., LEE, S.H., KO, G.I., KIM, J.B. and SOHN, D.H., 1996, Percutaneous absorption of antisense phosphorothioate oligonucleotide *in vitro*, *Arch. Pharm. Research.*, **19**, 116–121.

LELAWONGS, P., LIU, J., SIDDIQUI, O. and CHIEN, Y.W., 1989, Transdermal iontophoretic delivery of arginine-vasopressin (I): Physicochemical considerations, *Int. J. Pharm.*, **56**, 13–22.

LELAWONGS, P., LIU, J. and CHIEN, Y.W., 1990, Transdermal iontophoretic delivery of arginine-vasopressin (II): Evaluation of electrical and operational factors, *Int. J. Pharm.*, **61**, 179–188.

LEOPOLD, C.S. and LIPPOLD, B.C., 1992, A new application chamber for skin penetration studies *in vivo* with liquid preparations, *Pharm. Res.*, **9**, 1215–1218.

LEOPOLD, R.A., HUGHES, K.J. and DE VAULT, J.D., 1996, Using electroporation and a slot cuvette to deliver plasmid DNA to insect embryos, *Genet. Anal. Biomol. Eng.*, **12**, 197–200.

LEVIT, F., 1980, Treatment of hyperhidrosis by tap water iontophoresis, *Cutis*, **26**, 192–194.

LEWIS, K.J., IRWIN, W.J. and AKHTAR, S., 1995, Biodegradable poly(L-lactic acid) matrices for the sustained delivery of antisense oligonucleotides, *J. Control. Release*, **37**, 173–183.

LI, H. and KURAMITSU, H.K., 1996, Development of a gene transfer system in *Treponema denticola* by electroporation, *Oral Microbiol. Immunol.*, **11**, 161–165.

LI, L.C., SCUDDS, R.A., HECK, C.S. and HARTH, M., 1996, The efficacy of dexamethasone iontophoresis for the treatment of rheumatoid arthritic knees: A pilot study, *Arthritis Care Res.*, **9**, 126–132.

LI, S.K., GHANEM, A.H., PECK, K.D. and HIGUCHI, W.I., 1997, Iontophoretic transport across a synthetic membrane and human epidermal membrane: A study of the effects of permeant charge, *J. Pharm. Sci.*, **86**, 680–689.

LI, Z.J., JARRET, R.L., CHENG, M. and DEMSKI, J.W., 1995, Improved electroporation buffer enhances transient gene expression in *Arachis hypogaea* protoplasts, *Genome*, **38**, 858–863.

LIDEN, S., 1996, 'Sensitivity to electricity' – A new environmental epidemic, *Allergy*, **51**, 519–524.

LIN, R.Y., HSU, C.W. and CHEN, W.Y., 1996, A method to predict the transdermal permeability of amino acids and dipeptides through porcine skin, *J. Control. Release*, **38**, 229–234.

LIN, R.Y., OU, Y.C. and CHEN, W.Y., 1997, The role of electro-osmotic flow on *in vitro* transdermal iontophoresis, *J. Control. Release*, **43**, 23–33.

LIN, S.S., CHIEN, Y.W., HUANG, W.C., LI, C.H., CHUEH, C.L., CHEN, R.R.L., HSU, T.M., JIANG, T.S., WU, J.L. and VALIA, K.H. 1993a, Transdermal nicotine delivery systems – Multi-institutional cooperative bioequivalence studies, *Drug Dev. Ind. Pharm.*, **19**, 2765–2793.

LIN, S.S., HO, H. and CHIEN, Y.W., 1993b, Development of a new nicotine transdermal delivery system – *In vitro* kinetics studies and clinical pharmacokinetic evaluations in two ethnic groups, *J. Control. Release.*, **26**, 175–193.

LINBLAD, L.E. and EKENVALL, L., 1987, Electrode material in iontophoresis, *Pharm. Res.*, **4**, 438

LISS, H. and LISS, D., 1995, Lateral epicondylitis, *New Jersey Rehab.*, November, 8–11.

LIU, C.H., HO, H.O., HSIEH, M.C., SOKOLOSKI, T.D. and SHEU, M.T., 1995, Studies on the *in vitro* percutaneous penetration of indomethacin from gel systems in hairless mice, *J. Pharm. Pharmacol.*, **47**, 365–372.

LIU, J.C., SUN, Y., SIDDIQUI, O. and CHIEN, Y.W., 1988, Blood glucose control in diabetic rats by transdermal iontophoretic delivery of insulin, *Int. J. Pharm.*, **44**, 197–204.

LIU, P.C., NIGHTINGALE, J.A.S. and KURIHARABERGSTROM, T., 1993, Variation of human skin permeation *in vitro* – Ionic vs neutral compounds, *Int. J. Pharm.*, **90**, 171–176.

LOFTSSON, T., FRIORIKSDOTTIR, H., INGVARSDOTTIR, G., JONSDOTTIR, B. and SIGUROARDOTTIR, A.M., 1994, The influence of 2-hydroxypropyl-beta-cyclodextrin on diffusion rates and transdermal delivery of hydrocortisone, *Drug Dev. Ind. Pharm.*, **20**, 1699–1708.

LU, M.F., LEE, D. and RAO, G.S., 1992, Percutaneous absorption enhancement of leuprolide, *Pharm. Res.*, **9**, 1575–1579.

LU, S., ARTHOS, J., MONTEFIORI, D.C., YASUTOMI, Y., MANSON, K., MUSTAFA, F., JOHNSON, E., SANTORO, J.C., WISSINK, J., MULLINS, J.I., HAYNES, J.R., LETVIN, N.L., WYAND, M. and ROBINSON, H.L., 1996, Simian immunodeficiency virus DNA vaccine trial in macaques, *J. Virol.*, **70**, 3978–3991.

LUCA, A., GARCIAPAGAN, J.C., FEU, F., LOPEZTALAVERA, J.C., FERNANDEZ, M., BRU, C., BOSCH, J. and RODES, J., 1995, Noninvasive measurement of femoral blood flow and portal pressure response to propranolol in patients with cirrhosis, *Hepatology*, **21**, 83–88.

MACALUSO, R.A. and KENNEDY, T.L., 1989, Antibiotic iontophoresis in the treatment of burn perichondritis of the rabbit ear, *Otolaryngol. Head Neck Surg.*, **100**, 568–572.

MACK, K. and TITBALL, R.W., 1996, Transformation of *Burkholderia pseudomallei* by electroporation, *Anal. Biochem.*, **242**, 73–76.
MAHATO, R.I., ROLLAND, A. and TOMLINSON, E., 1997, Cationic lipid-based gene delivery systems: Pharmaceutical perspectives, *Pharm. Res.*, **14**, 853–859.
MAITANI, Y., COUTELEGROS, A., OBATA, Y. and NAGAI, T., 1993, Prediction of skin permeabilities of diclofenac and propranolol from theoretical partition coefficients determined from cohesion parameters, *J. Pharm. Sci.*, **82**, 416–420.
MARTTIN, E., NEELISSEN SUBNEL, M.T.A., DE HAAN, F.H.N. and BODDE, H.E., 1996, A critical comparison of methods to quantify stratum corneum removed by tape stripping, *Skin Pharmacol.*, **9**, 69–77.
MASADA, T., HIGUCHI, W.I., SRINIVASAN, V., ROHR, U., FOX, J., BEHL, C. and PONS, S., 1989, Examination of iontophoretic transport of ionic drugs across skin: Baseline studies with the four-electrode system, *Int. J. Pharm.*, **49**, 57–62.
MATTHEWS, K.E., DEV, S.B., TONEGUZZO, F. and KEATING, A., 1995, Electroporation for gene therapy, *Animal Cell Electroporation Electrofusion Protocols*, **48**, 280.
MATUSZEWSKA, B., KEOGAN, M., FISHER, D.M., SOPER, K.A., HOE, C., HUBER, A.C. and BONDI, J.V., 1994, Acidic fibroblast growth factor: Evaluation of topical formulations in a diabetic mouse wound healing model, *Pharm. Res.*, **11 (1)**, 65–71.
MAURER, D.D., WILLIAMS, T.J. and STEVENS, S.A., 1993, Multiple site drug iontophoresis electronic device and method, U.S. Pat. 5 254 081. Assignee: Empi.
MAURICE, D.M., 1986, Iontophoresis of fluorescein into the posterior segment of the rabbit eye, *Ophthalmology*, **93**, 128–132.
MEIDAN, V.M., WALMSLEY, A.D. and IRWIN, W.J., 1995, Phonophoresis – Is it a reality? *Int. J. Pharm.*, **118**, 129–149.
MELIKOV, K.C. and ERSHLER, I.A., 1996, Localization of conductive pathways in human stratum corneum by scanning electrochemical microscopy, *Membr. Cell Biol.*, **10**, 459–466.
MELKONYAN, H., SORG, C. and KLEMPT, M., 1996, Electroporation efficiency in mammalian cells is increased by dimethyl sulfoxide (DMSO), *Nucleic Acids Res.*, **24**, 4356–4357.
MELSKI, J.W., 1996, The anatomy and physiology of the skin, in: *Principles and Practice of Skin Excisions*, B.I.E. JEMEC and G.B.E. JEMEC (eds). Elsevier, Amsterdam, pp. 1–14.
MENASCHE, M., JACOB, M.P., GODEAU, G., ROBERT, A.M. and ROBERT, L., 1981, Pharmacological studies on elastin peptides (kappa-elastin): Blood clearance, percutaneous penetration and tissue distribution, *Path. Biol.*, **29**, 548–554.
MERINO, V., KALIA, Y.N. and GUY, R.H., 1997, Transdermal therapy and diagnosis by iontophoresis, *Trends. Biotech.*, **15**, 288–290.
MERK, H.F., JUGERT, F.K. and FRANKENBERG, S., 1996, Biotransformations in the skin, in: *Dermatotoxicology*, F.N. MARZULLI and H.I. MAIBACH (eds). Taylor & Francis, USA, pp. 61–73.
MEYER, B.R., KREIS, W. and ESCHBACH, J., 1988, Successful transdermal administration of therapeutic doses of a polypeptide to normal human volunteers, *Clin. Pharmacol. Ther.*, **44**, 607–612.
MEYER, B.R., KATZEFF, H.L., ESCHBACH, J.C., TRIMMER, J., ZACHARIAS, S.B., ROSEN, S. and SIBALIS, D., 1989, Transdermal delivery of human insulin to albino rabbits using electrical current, *Am. J. Med. Sci.*, **297**, 321–325.
MEYER, B.R., KREIS, W., ESCHBACH, J., O'MARA, V., ROSEN, S. and SIBALIS, D., 1990, Transdermal versus subcutaneous leuprolide: A comparison of acute pharmacodynamic effect, *Clin. Pharmacol. Ther.*, **48**, 340–345.
MICHAELS, A.S., CHANDRASEKARAN, S.K. and SHAW, J.E., 1975, Drug permeation through human skin: Theory and *in vitro* experimental measurement, *AICHE J.*, **21**, 985–996.
MICHNIAK, B.B., PLAYER, M.R., CHAPMAN, J.M. and SOWELL, J.W., 1994, Azone analogues as penetration enhancers: Effect of different vehicles on hydrocortisone acetate skin permeation and retention, *J. Control. Release*, **32**, 147–154.
MIDTGAARD, K., 1986, A new device for the treatment of hyperhidrosis by iontophoresis, *Br. J. Dermatol.*, **114**, 485–488.

MIGUEL, R., KREITZER, J.M., REINHART, D., SEBEL, P.S., BOWIE, J., FREEDMAN, G. and EISENKRAFT, J.B., 1995, Postoperative pain control with a new transdermal fentanyl delivery system: A multicenter trial, *Anesthesiology*, **83**, 470–477.

MIKLAVCIC, D., SEMROV, D., VALENCIC, V., SERSA, G. and VODOVNIK, L., 1997, Tumor treatment by direct electric current: Computation of electric current and power density distribution, *Electro. Magnetobiol.*, **16**, 119–128.

MILLER, L.L. and SMITH, G.A., 1989, Iontophoretic transport of acetate and carboxylate ions through hairless mouse skin. A cation exchange membrane model, *Int. J. Pharm.*, **49**, 15–22.

MILLER, L.L., SMITH, G.A., CHANG, A. and ZHOU, Q., 1987, Electrochemically controlled release, *J. Control. Release*, **6**, 293–296.

MILLER, L.L., KOLASKIE, C.J., SMITH, G.A. and RIVIER, J., 1990, Transdermal iontophoresis of gonadotropin releasing hormone (LHRH) and two analogues, *J. Pharm. Sci.*, **79**, 490–493.

MITRAGOTRI, S., BLANKSCHTEIN, D. and LANGER, R. 1995a, Ultrasound-mediated transdermal protein delivery, *Science*, **269**, 850–853.

MITRAGOTRI, S., EDWARDS, D.A., BLANKSCHTEIN, D. and LANGER, R., 1995b, A mechanistic study of ultrasonically-enhanced transdermal drug delivery, *J. Pharm. Sci.*, **84**, 697–706.

MOLL, F. and KNOBLAUCH, P., 1993, Iontophoretic *in vitro* release of antimycotics from hydrogels, *Drug Dev. Ind. Pharm.*, **19**, 1143–1158.

MOLONEY, P.J., APRILE, M.A. and WILSON, S., 1964, Sulfated insulin for treatment of insulin-resistant diabetics, *J. New Drugs*, September/October, 258–263.

MONTEIRO-RIVIERE, N.A., 1990, Altered epidermal morphology secondary to lidocaine iontophoresis: *In vivo* and *in vitro* studies in porcine skin, *Fundam. Appl. Toxicol.*, **15**, 174–185.

MONTEIRO-RIVIERE, N.A., INMAN, A.O. and RIVIERE, J.E., 1994, Identification of the pathway of iontophoretic drug delivery: Light and ultrastructural studies using mercuric chloride in pigs, *Pharm. Res.*, **11(2)**, 251–256.

MOODY, R.P., 1997, Automated *in vitro* dermal absorption (AIVDA): A new *in vitro* method for investigating transdermal flux, *ATLA*, **25**, 347–357.

MORIMOTO, K., IWAKURA, Y., NAKATANI, E., MIYAZAKI, M. and TOJIMA, H., 1992, Effects of proteolytic enzyme inhibitors as absorption enhancers on the transdermal iontophoretic delivery of calcitonin in rats, *J. Pharm. Pharmacol.*, **44**, 216–218.

MORLEY, S.M., 1997, Keratin and the skin: Past, present and future, *Q. J. Med.*, **90**, 433–435.

MORRIS, S.J. and SHORE, A.C., 1996, Skin blood flow responses to the iontophoresis of acetylcholine and sodium nitroprusside in man: Possible mechanisms, *J. Physiol.*, **496(2)**, 531–542.

MOSEKILDE, E., JENSEN, K.S., BINDER, C., PRAMMING, S. and THORSTEINSSON, B., 1989, Modeling absorption kinetics of subcutaneous injected soluble insulin, *J. Pharmacokinet. Biopharm.*, **17**, 67–87.

MOSS, J. and BUNDGAARD, H., 1990, Prodrugs of Peptides. 7. Transdermal delivery of thyrotropin-releasing hormone (TRH) via prodrugs, *Int. J. Pharm.*, **66**, 39–45.

MUDDLE, A.G., LONGRIDGE, D.J., SWEENEY, P.A., BURKOTH, T.L. and BELLHOUSE, B.J., 1997, Transdermal delivery of testosterone to conscious rabbits using Powderject® – a supersonic powder delivery system., *Proc. Int'l Symp. Control. Rel. Bioact. Mater.*, Controlled Release Society, Inc., 24. pp. 713–714.

MULLER, M., SCHMID, R., WAGNER, O., VONOSTEN, B., SHAYGANFAR, H. and EICHLER, H.G., 1995, *In vivo* characterization of transdermal drug transport by microdialysis, *J. Control. Release*, **37**, 49–57.

NAEFF, R., 1996, Feasibility of topical liposome drugs produced on an industrial scale, *Adv. Drug Del. Rev.*, **18**, 343–347.

NAKAKURA, M., TERAJIMA, M., KATO, Y., HAYAKAWA, E., ITO, K. and KURODA, T., 1995, Effect of iontophoretic patterns on *in vivo* antidiuretic response to desmopressin acetate administered transdermally, *J. Drug Target.*, **2**, 487–492.

NAKAKURA, M., KATO, Y., HAYAKAWA, E., ITO, K. and KURODA, T., 1996, Effect of pulse on iontophoretic delivery of desmopressin acetate in rats, *Biol. Pharm. Bull.*, **19**, 738–740.

NAKAKURA, M., KATO, Y. and ITO, K., 1997, Prolongation of antidiuretic response to desmopressin acetate by iontophoretic transdermal delivery in rats, *Biol. Pharm. Bull.*, **20**, 537–540.

NAKHARE, S., JAIN, N.K. and VERMA, H.V., 1994, Iontophoretic cellophane membrane delivery of diclofenac sodium, *Pharmazie*, **49**, 672–675.

NANDA, A. and KHAR, R.K., 1994, Enhancement of percutaneous absorption of propranolol hydrochloride by iontophoresis, *Drug Develop. Ind. Pharm.*, **20**, 3033–3044.

NATSUKI, R., MORITA, Y., OSAWA, S. and TAKEDA, Y., 1996, Effects of liposome size on penetration of dl-tocopherol acetate into skin, *Biol. Pharm. Bull.*, **19**, 758–761.

NEUMANN, E. 1996, Gene delivery by membrane electroporation, in: *Electrical Manipulation of Cells*, P.T. LYNCH and M.R. DAVEY (eds). Chapman & Hall, pp. 157–183.

NEUMANN, E., KAKORIN, S., TSONEVA, I., NIKOLOVA, B. and TOMOV, T., 1996, Calcium-mediated DNA adsorption to yeast cells and kinetics of cell transformation by electroporation, *Biophys. J.*, **71**, 868–877.

NOLEN, H.W., CATZ, P. and FRIEND, D.R., 1994, Percutaneous penetration of methyl phosphonate antisense oligonucleotides, *Int. J. Pharm.*, **107**, 169–177.

NUMAJIRI, S., SUGIBAYASHI, K. and MORIMOTO, Y., 1993, Non-invasive sampling of lactic acid ions by iontophoresis using chloride ion in the body as an internal standard, *J. Pharm. Biomed. Anal.*, **11**, 903–909.

O'MALLEY, E.P. and OESTER, Y.T., 1955, Influence of some physical chemical factors on iontophoresis using radio isotopes, *Arch. Phys. Med. Rehabil.*, **36**, 310–316.

OBATA, Y., TAKAYAMA, K., MAITANI, Y., MACHIDA, Y. and NAGAI, T., 1993, Effect of ethanol on skin permeation of nonionized and ionized diclofenac, *Int. J. Pharm.*, **89**, 191–198.

OGISO, T. and SHINTANI, M., 1990, Mechanism for the enhancement effect of fatty acids on the percutaneous absorption of propranolol, *J. Pharm. Sci.*, **79**, 1065–1071.

OGISO, T., NISHIOKA, S. and IWAKI, M., 1996, Dissociation of insulin oligomers and enhancement of percutaneous absorption of insulin, *Biol. Pharm. Bull.*, **19**, 1049–1054.

OH, S.Y. and GUY, R.H., 1995, Effects of iontophoresis on the electrical properties of human skin *in vivo*, *Int. J. Pharm.*, **124**, 137–142.

OH, S.Y., LEUNG, L., BOMMANNAN, D., GUY, R.H. and POTTS, R.O., 1993, Effect of current, ionic strength and temperature on the electrical properties of skin, *J. Control. Release*, **27**, 115–125.

OH, S.Y., JEONG, S.Y., and LEE, J.H., 1997, Iontophoretic delivery of azidothymidine, *Proc. Int'l Symp. Control. Rel. Bioact. Mater.*, Controlled Release Society, Inc., 24. pp. 697–698.

OKABE, K., YAMAGUCHI, H. and KAWAI, Y., 1986, New iontophoretic transdermal administration of the beta blocker metoprolol, *J. Control. Release*, **4**, 79–85.

OKUMURA, K., KIYOHARA, Y., KOMADA, F., IWAKAWA, S., HIRAI, M. and FUWA, T., 1990, Improvement in wound healing by epidermal growth factor (EGF) ointment. I. Effect of nafamostat, gabexate, or gelatin on stabilization and efficacy of EGF, *Pharm. Res.*, **7**, 1289–1293.

OKUMURA, M., OKUDA, T., NAKAMURA, T. and YAJIMA, M., 1996, Acceleration of wound healing in diabetic mice by basic fibroblast growth factor, *Biol. Pharm. Bull.*, **19**, 530–535.

OLDENBURG, K.R., VO, K.T., SMITH, G.A. and SELICK, H.E., 1995, Iontophoretic delivery of oligonucleotides across full thickness hairless mouse skin, *J. Pharm. Sci.*, **84**, 915–921.

OLLMAR, S. and EMTESTAM, L., 1992, Electrical impedance applied to noninvasive detection of irritation in skin, *Contact Dermatitis*, **27**, 37–42.

OLLMAR, S., NYREN, M., NICANDER, I. and EMTESTAM, L., 1994, Electrical impedance compared with other non-invasive bioengineering techniques and visual scoring for detection of irritation in human skin, *Br. J. Dermatol.*, **130**, 29–36.

OPHIR, J., BRENNER, S., BALI, R., KRISSLEVENTON, S., SMETANA, Z. and REVEL, M., 1995, Effect of topical interferon-beta on recurrence rates in genital herpes: A double-blind, placebo-controlled, randomized study, *J. Interferon Cytokine Res.*, **15**, 625–631.

ORSINI, J.C., BARONE, F.C., ARMSTRONG, D.L. and WAYNER, M.J., 1985, Direct effects of androgens on lateral hypothalamic neuronal activity in the male rat I. A microiontophoretic study, *Brain Res. Bull.*, **15**, 293–297.

PANUS, P. and BANGA, A.K., 1997, Iontophoresis devices: Clinical applications for topical delivery, *Int. J. Pharm. Compd.*, **1**, 420–424.

PANUS, P.C., KULKARNI, S.B., CAMPBELL, J., RAVIS, W.R. and BANGA, A.K., 1996, Effect of iontophoretic current and application time on transdermal delivery of ketoprofen in man, *Pharm. Sci.*, **2**, 467–469.

PANUS, P.C., CAMPBELL, J., KULKARNI, S.B., HERRICK, R.T., RAVIS, W.R. and BANGA, A.K., 1997, Transdermal iontophoretic delivery of ketoprofen through human cadaver skin and in humans, *J. Control. Release*, **44**, 113–121.

PARASRAMPURIA, D. and PARASRAMPURIA, J., 1991, Percutaneous delivery of proteins and peptides using iontophoretic techniques, *J. Clin. Pharm. Ther.*, **16**, 7–17.

PARK, N.H., GANGAROSA, L.P., KWON, B.S. and HILL, J.M., 1978, Iontophoretic application of adenosine arabinoside monophosphate to herpes simplex virus type 1 infected hairless mouse skin, *Antimicrob. Agents Chemother.*, **14**, 605–608.

PAUL, A., CEVC, G. and BACHHAWAT, B.K., 1995, Transdermal immunization with large proteins by means of ultradeformable drug carriers, *Eur. J. Immunol.*, **25**, 3521–3524.

PAUS, R., 1997, Immunology of the hair follicle, in: *Skin Immune System: Cutaneous immunology and clinical immunodermatology*, J.D. Bos (ed.), 2nd edn. CRC Press, Boca Raton, pp. 377–398.

PAYNE, R., CHANDLER, S. and EINHAUS, M., 1995, Guidelines for the clinical use of transdermal fentanyl, *Anti-Cancer Drug*, **6**, 50–53.

PECHTOLD, L.A.R.M., ABRAHAM, W. and POTTS, R.O., 1996, The influence of an electric field on ion and water accessibility to stratum corneum lipid lamellae, *Pharm. Res.*, **13**, 1168–1173.

PECK, K.D., GHANEM, A.H., HIGUCHI, W.I. and SRINIVASAN, V., 1993, Improved stability of the human epidermal membrane during successive permeability experiments, *Int. J. Pharm.*, **98**, 141–147.

PECK, K.D., SRINIVASAN, V., LI, S.K., HIGUCHI, W.I. and GHANEM, A.H., 1996, Quantitative description of the effect of molecular size upon electro-osmotic flux enhancement during iontophoresis for a synthetic membrane and human epidermal membrane, *J. Pharm. Sci.*, **85**, 781–788.

PELLECCHIA, G.L., HAMEL, H. and BEHNKE, P., 1994, Treatment of infrapatellar tendinitis: A combination of modalities and transverse friction message versus iontophoresis, *J. Sport Rehab.*, **3**, 135–145.

PETELENZ, T., AXENTI, I., PETELENZ, T.J., IWINSKI, J. and DUBEL, S., 1984, Mini set for iontophoresis for topical analgesia before injection, *J. Clin. Pharmacol. Ther. Toxicol.*, **22**, 152–155.

PETELENZ, T.J., JACOBSEN, S.C., STEPHEN, R.L. and JANATA, J., 1990, Methods and apparatus for iontophoresis applications of medicaments at a controlled pH through ion exchange, U.S. Pat. 4915685.

PETELENZ, T.J., BUTTKE, J.A., BONDS, C., LLOYD, L.B., BECK, J.E., STEPHEN, R.L., JACOBSEN, S.C. and RODRIGUEZ, P., 1992, Iontophoresis of dexamethasone: laboratory studies, *J. Control. Release*, **20**, 55–66.

PETERSON, J.L., READ, S.I. and RODMAN, O.G., 1982, A new device in the treatment of hyperhidrosis by iontophoresis, *Cutis*, **29**, 82–89.

PFISTER, W.R. and HSIEH, D.S.T., 1990a, Permeation enhancers compatible with transdermal drug delivery systems: Part II: System design considerations, *Pharm. Technol.*, **14**(10), 54–60.

1990b, Permeation enhancers compatible with transdermal drug delivery systems. Part I: Selection and formulation considerations, *Pharm. Technol.*, **14**(9), 132–140.

PHIPPS, J.B., 1995, Method for reducing sensation in iontophoretic drug delivery, U.S. Pat. 5403275. Assignee: Alza Corp.

PHIPPS, J.B. and GYORY, J.R., 1992, Transdermal ion migration, *Adv. Drug Del. Rev.*, **9**, 137–176.

PHIPPS, J.B., PADMANABHAN, R.V. and LATTIN, G.A., 1989, Iontophoretic delivery of model inorganic and drug ions, *J. Pharm. Sci.*, **78**, 365–369.

PHIPPS, J.B., HAAK, R., CHAO, S., GUPTA, S.K. and GYORY, J.R., 1994, In vivo transdermal electrotransport in humans, *Proc. Int'l Symp. Control. Rel. Bioact. Mater.*, Controlled Release Society, Inc., 21, pp. 170–171.

PIKAL, M.J., 1990, Transport mechanisms in iontophoresis. I. A theoretical model for the effect of electro-osmotic flow on flux enhancement in transdermal iontophoresis, *Pharm. Res.*, **7**, 118–126.

1992, The role of electro-osmotic flow in transdermal iontophoresis, *Adv. Drug Del. Rev.*, **9**, 201–237.

1995, Penetration enhancement of peptide and protein drugs by electrochemical means: Transdermal iontophoresis, in: *Trends and Future Perspectives in Peptide and Protein Drug Delivery*, V.H.L. LEE, M. HASHIDA, and Y. MIZUSHIMA, (eds). Harwood Academic Publishers GmbH, Chur, Switzerland, pp. 83–109.

PIKAL, M.J. and SHAH, S., 1990a, Transport mechanisms in iontophoresis. II. Electroosmotic flow and transference number measurements for hairless mouse skin, *Pharm. Res.*, **7**, 213–221.

1990b, Transport mechanisms in iontophoresis. III. An experimental study of the contributions of electro-osmotic flow and permeability change in transport of low and high molecular weight solutes, *Pharm. Res.*, **7**, 222–229.

1991, Study of the mechanisms of flux enhancement through hairless mouse skin by pulsed DC iontophoresis, *Pharm. Res.*, **8**, 365–369.

PITMAN, I.H. and ROSTAS, S.J., 1981, Topical drug delivery to cattle and sheep, *J. Pharm. Sci.*, **70**, 1181–1194.

PIZZAMIGLIO, L., VALLAR, G. and MAGNOTTI, L., 1996, Transcutaneous electrical stimulation of the neck muscles and hemineglect rehabilitation, *Restor. Neurol. Neurosci.*, **10**, 197–203.

PLIQUETT, F. and PLIQUETT, U., 1996, Passive electrical properties of human stratum corneum *in vitro* depending on time after separation, *Biophys. Chem.*, **58**, 205–210.

PLIQUETT, U. and WEAVER, J.C., 1996a, Electroporation of human skin: Simultaneous measurement of changes in the transport of two fluorescent molecules and in the passive electrical properties, *Bioelectrochem. Bioenerg.*, **39**, 1–12.

1996b, Transport of a charged molecule across the human epidermis due to electroporation, *J. Control. Release*, **38**, 1–10.

PLIQUETT, U., LANGER, R. and WEAVER, J.C., 1995a, Changes in the passive electrical properties of human stratum corneum due to electroporation, *Biochim. Biophys. Acta*, **1239**, 111–121.

PLIQUETT, U., PRAUSNITZ, M.R., CHIZMADZHEV, Y.A. and WEAVER, J.C., 1995b, Measurement of rapid release kinetics for drug delivery, *Pharm. Res.*, **12**, 549–555.

PLIQUETT, U., GIFT, E.A. and WEAVER, J.C., 1996, Determination of the electric field and anomalous heating caused by exponential pulses with aluminum electrodes in electroporation experiments, *Bioelectrochem. Bioenerg.*, **39**, 39–53.

PLUTCHIK, R. and HIRSCH, H.R., 1963, Skin impedance and phase angle as a function of frequency and current, *Science*, **141**, 927–928.

PORTELLA, G., LIDDELL, J., CROMBIE, R., HADDOW, S., CLARKE, M., STOLER, A.B. and BALMAIN, A., 1994, Molecular mechanisms of invasion and metastasis during mouse skin tumour progression, *Invasion Metastasis*, **14**, 7–16.

POTTS, R.O. and CLEARY, G.W., 1995, Transdermal drug delivery: Useful paradigms, *J. Drug Target.*, **3**, 247–251.

POTTS, R.O. and GUY, R.H., 1992, Predicting skin permeability, *Pharm. Res.*, **9**, 663–669.

POTTS, R.O., MCNEILL, S.C., DESBONNET, C.R. and WAKSHULL, E., 1989, Transdermal drug transport and metabolism. II. The role of competing kinetic events, *Pharm. Res.*, **6**, 119–124.

POTTS, R.O., BOMMANNAN, D., WONG, O., TAMADA, J.A., RIVIERE, J.E. and MONTEIRORIVIERE, N.A., 1997, Transdermal peptide delivery using electroporation, in: *Protein Delivery: Physical Systems*, Sanders and Hendren (eds). Plenum Press, New York, pp. 213–238.

POULOS, C.W., BRENNER, G.M., ALLEN, L.V., PRABHU, V.A. and HUERTA, P.L., 1995, Method for stimulating hair growth with cationic derivative of minoxidil using therapeutic iontophoresis, U.S. Pat. 5466695.

POWERS, W.R. and SISUN, H., 1987, Electrodes, electrode assemblies, methods, and systems for tissue stimulation and transdermal delivery of pharmacologically active ligands, U.S. Pat. 4702732. Assignee: Boston Univ.

1988, Electrodes, electrode assemblies, methods, and systems for tissue stimulation, U.S. Pat. 4786277. Assignee: Boston Univ.

PRATHER, R.D., TU, T.G., ROLF, C.N. and GORSLINE, J., 1993, Nicotine pharmacokinetics of nicoderm (R) (Nicotine transdermal system) in women and obese men compared with normal-sized men, *J. Clin. Pharmacol.*, **33**, 644–649.

PRATZEL, H., DITTRICH, P. and KUKOVETZ, W., 1986, Spontaneous and forced cutaneous absorption of indomethacin in pigs and humans, *J. Rheumatol.*, **13**, 1122–1125.

PRAUSNITZ, M.R., 1996a, The effects of electric current applied to skin: A review for transdermal drug delivery, *Adv. Drug Del. Rev.*, **18**, 395–425.

1996b, Do high-voltage pulses cause changes in skin structure? *J. Control. Release*, **40**, 321–326.

1997, The effect of pulsed electrical protocols on skin damage, sensation and pain, *Proc. Int'l Symp. Control. Rel. Bioact. Mater.*, Controlled Release Society, Inc., 24, pp. 25–26.

PRAUSNITZ, M.R., BOSE, V.G., LANGER, R. and WEAVER, J.C., 1993a, Electroporation of mammalian skin – A mechanism to enhance transdermal drug delivery, *Proc. Natl. Acad. Sci. USA*, **90**, 10 504–10 508.

PRAUSNITZ, M.R., SEDDICK, D.S., KON, A.A., BOSE, V.G., FRANKENBURG, S., KLAUS, S.N., LANGER, R. and WEAVER, J.C., 1993b, Methods for *in vivo* tissue electroporation using surface electrodes, *Drug Del.*, **1**, 125–131.

PRAUSNITZ, M.R., PLIQUETT, U., LANGER, R. and WEAVER, J.C., 1994, Rapid temporal control of transdermal drug delivery by electroporation, *Pharm. Res.*, **11**, 1834–1837.

PRAUSNITZ, M.R., CORBETT, J.D., GIMM, J.A., GOLAN, D.E., LANGER, R. and WEAVER, J.C., 1995a, Millisecond measurement of transport during and after an electroporation pulse, *Biophys. J.*, **68**, 1864–1870.

PRAUSNITZ, M.R., EDELMAN, E.R., GIMM, J.A., LANGER, R. and WEAVER, J.C., 1995b, Transdermal delivery of heparin by skin electroporation, *Biotechnology*, **13**, 1205–1209.

PRAUSNITZ, M.R., GIMM, J.A., GUY, R.H., LANGER, R., WEAVER, J.C. and CULLANDER, C., 1996a, Imaging regions of transport across human stratum corneum during high-voltage and low-voltage exposures, *J. Pharm. Sci.*, **85**, 1363–1370.

PRAUSNITZ, M.R., LEE, C.S., LIU, C.H., PANG, J.C., SINGH, T.P., LANGER, R. and WEAVER, J.C., 1996b, Transdermal transport efficiency during skin electroporation and iontophoresis, *J. Control. Release*, **38**, 205–217.

PREAT, V. and THYSMAN, S., 1993, Transdermal ionotophoretic delivery of sufentanil, *Int. J. Pharm.*, **96**, 189–196.

PREISS, A., MEHNERT, W. and FROMMING, K.H., 1995, Penetration of hydrocortisone into excised human skin under the influence of cyclodextrins, *Pharmazie*, **50**, 121–126.

PUGH, W.J. and HADGRAFT, J., 1994, *Ab initio* prediction of human skin permeability coefficients, *Int. J. Pharm.*, **103**, 163–178.

PUTTEMANS, F.J.M., MASSART, D.L., GILES, F., LIEVENS, P.C. and JONCKEER, M.H., 1982, Iontophoresis: Mechanism of action studied by potentiometry and X-ray fluorescence, *Arch. Phys. Med. Rehabil.*, **63**, 176–180.

QIAO, G.L. and RIVIERE, J.E., 1995, Significant effects of application site and occlusion on the pharmacokinetics of cutaneous penetration and biotransformation of parathion *in vivo* in swine, *J. Pharm. Sci.*, **84**, 425–432.

QIN, X.S., ZHANG, S.M., ODA, K., NAKATSURU, Y., SHIMIZU, S., YAMAZAKI, Y., NIKAIDO, O. and ISHIKAWA, T., 1995, Quantitative detection of ultraviolet light-induced photoproducts in mouse skin by immunohistochemistry, *Jpn. J. Cancer Res.*, **86**, 1041–1048.

RADOMSKA, H.S. and ECKHARDT, L.A., 1995, Mammalian cell fusion in an electroporation device, *J. Immunol. Methods*, **188**, 209–217.

RAMSDEN, R.T., GIBSON, W.P.R. and MOFFAT, D.A., 1977, Anesthesia of the tympanic membrane using iontophoresis, *J. Laryngol. Otol.*, **91**, 779–785.

RANADE, V.V. 1991, Drug delivery systems. 6. Transdermal drug delivery, *J. Clin. Pharmacol.*, **31**, 401–418.

Rao, G., Glikfeld, P. and Guy, R.H., 1993, Reverse iontophoresis: Development of a noninvasive approach for glucose monitoring, *Pharm. Res.*, **10**, 1751–1755.

Rao, G., Guy, R.H., Glikfeld, P., Lacourse, W.R., Leung, L., Tamada, J., Potts, R.O. and Azimi, N., 1995, Reverse iontophoresis: Noninvasive glucose monitoring *in vivo* in humans, *Pharm. Res.*, **12**, 1869–1873.

Rao, V.U. and Misra, A.N., 1994, Enhancement of iontophoretic permeation of insulin across human cadaver skin, *Pharmazie*, **49**, 538–539.

Rapperport, A.S., Larson, D.L., Henges, D.F., Lynch, J.B., Blocker, T.G. and Lewis, R.S., 1965, Iontophoresis: A method of antibiotic administration in the burn patient, *Plastic Reconstruc. Surg.*, **36**, 547–552.

Rattenbury, J.M. and Worthy, E., 1996, Is the sweat test safe? Some instances of burns received during pilocarpine iontophoresis, *Ann. Clin. Biochem.*, **33**, 456–458.

Raykar, P.V., Fung, M. and Anderson, B.D., 1988, The role of protein and lipid domains in the uptake of solutes by human stratum corneum, *Pharm. Res.*, **5**, 140–150.

Raz, E., Carson, D.A., Parker, S.E., Parr, T.B., Abai, A.M., Aichinger, G., Gromkowski, S.H., Singh, M., Lew, D., Yankauckas, M.A., Baird, S.M. and Rhodes, G.H., 1994, Intradermal gene immunization: The possible role of DNA uptake in the induction of cellular immunity to viruses, *Proc. Natl. Acad. Sci. USA*, **91**, 9519–9523.

Regnier, V., Tahiri, A., Andre, N., Lamaitre, M., Preat, V. and Doan, T.L. 1997, Delivery of 3′-protected phosphodiester oligodeoxynucleotides to the skin, *Pharm. Res.*, **14**, S-640.

Reid, K.I., Dionne, R.A., Sicard-Rosenbaum, L., Lord, D. and Dubner, R.A., 1994, Evaluation of iontophoretically applied dexamethasone for painful pathologic temporomandibular joints, *Oral Surg. Oral Med. Oral Pathol.*, **77**, 605–609.

Reifenrath, W.G., Hawkins, G.S. and Kurtz, M.S., 1991, Percutaneous penetration and skin retention of topically applied compounds: An *in vitro*–*in vivo* study, *J. Pharm. Sci.*, **80**, 526–532.

Reinauer, S., Neusser, A., Schauf, G. and Holzle, E., 1993, Iontophoresis with alternating current and direct current offset (AC DC iontophoresis) – A new approach for the treatment of hyperhidrosis, *Br. J. Dermatol.*, **129**, 166–169.

Rigano, W., Yanik, M., Barone, F.A., Baibak, G. and Cislo, C., 1992, Antibiotic iontophoresis in the management of burned ears, *J. Burn Care Rehabil.*, **13**, 407–409.

Rittich, B. and Spanova, A., 1996, Electrotransformation of bacteria by plasmid DNAs: Statistical evaluation of a model quantitatively describing the relationship between the number of electrotransformants and DNA concentration, *Bioelectrochem. Bioenerg.*, **40**, 233–238.

Rittich, B., Manova, K., Spanova, A. and Pribyla, L., 1996, Electrotransformation of bacteria by plasmid DNA: Effect of serial electroporator resistor, *Chem. Papers*, **50**, 245–248.

Riviere, J.E., 1996, Isolated perfused porcine skin flap, in: *Dermatotoxicology*, F.N. Marzulli and H.I. Maibach, (eds). Taylor & Francis, USA, pp. 337–351.

Riviere, J.E. and Heit, M.C., 1997, Electrically-assisted transdermal drug delivery, *Pharm. Res.*, **14**, 687–697.

Riviere, J.E., Bowman, K.F., Monteiro-Riviere, N.A., Dix, L.P. and Carver, M.P., 1986, The isolated perfused porcine skin flap (IPPSF): I. A novel *in vitro* model for percutaneous absorption and cutaneous toxicology studies, *Fundam. Appl. Toxicol.*, **7**, 444–453.

Riviere, J.E., Sage, B. and Williams, P.L., 1991, Effects of vasoactive drugs on transdermal lidocaine iontophoresis, *J. Pharm. Sci.*, **80**, 615–620.

Riviere, J.E., Monteiro-Riviere, N.A. and Inman, A.O., 1992a, Determination of lidocaine concentrations in skin after transdermal iontophoresis: Effects of vasoactive drugs, *Pharm. Res.*, **9**, 211–214.

Riviere, J.E., Williams, P.L., Hillman, R.S. and Mishky, L.M., 1992b, Quantitative prediction of transdermal iontophoretic delivery of arbutamine in humans with the *in vitro* isolated perfused porcine skin flap, *J. Pharm. Sci.*, **81**, 504–507.

Riviere, J.E., Monteioriviere, N.A., Rogers, R.A., Bommannan, D., Tamada, J.A. and Potts, R.O., 1995, Pulsatile transdermal delivery of LHRH using electroporation: Drug delivery and skin toxicology, *J. Control. Release*, **36**, 229–233.

RIVIERE, J.E., BROOKS, J.D., WILLIAMS, P.L., MCGOWN, E. and FRANCOEUR, M.L., 1996, Cutaneous metabolism of isosorbide dinitrate after transdermal administration in isolated perfused porcine skin, *Int. J. Pharm.*, **127**, 213–217.

ROBINSON, H.L., LU, S., FELTQUATE, D.M., TORRES, C.T., RICHMOND, J., BOYLE, C.M., MORIN, M.J., SANTORO, J.C., WEBSTER, R.G., MONTEFIORI, D., YASUTOMI, Y., LETVIN, N.L., MANSON, K., WYAND, M. and HAYNES, J.R., 1996, DNA vaccines, *AIDS Res. Hum. Retroviruses*, **12**, 455–457.

ROLLAND, A., 1993, Particulate carriers in dermal and transdermal drug delivery: Myth or reality? in: *Pharmaceutical Particulate Carriers: Therapeutic applications*. A. Rolland, (ed.). Marcel Dekker, Inc., New York, pp. 367–421.

ROSEMBERG, Y. and KORENSTEIN, R., 1997, Incorporation of macromolecules into cells and vesicles by low electric fields: induction of endocytotic-like processes, *Bioelectrochem. Bioenerg.*, **42**, 275–281.

ROSEY, E.L., KENNEDY, M.J., PETRELLA, D.K., ULRICH, R.G. and YANCEY, R.J., 1995, Inactivation of *Serpulina hyodysenteriae flaA1* and *flaB1* periplasmic flagellar genes by electroporation mediated allelic exchange, *J. Bacteriol.*, **177**, 5959–5970.

ROTHFELD, S.H. and MURRAY, W., 1967, The treatment of Peyronie's disease by iontophoresis of C_{21} esterified glucocorticoids, *J. Urology*, **97**, 874–875.

ROY, S.D. and DEGROOT, J.S., 1994, Percutaneous absorption of nafarelin acetate, an LHRH analogue, through human cadaver skin and monkey skin, *Int. J. Pharm.*, **110**, 137–145.

ROY, S.D. and FLYNN, G.L., 1988, Solubility and related physicochemical properties of narcotic analgesics, *Pharm. Res.*, **5**, 580–586.

1990, Transdermal delivery of narcotic analgesics: pH, anatomical, and subject influences on cutaneous permeability of fentanyl and sufentanil, *Pharm. Res.*, **7**, 842–847.

ROY, S.D. and MANOUKIAN, E., 1994, Permeability of ketorolac acid and its ester analogues (prodrug) through human cadaver skin, *J. Pharm. Sci.*, **83**, 1548–1553.

ROY, S.D., HOU, S.Y.E., WITHAM, S.L. and FLYNN, G.L., 1994a, Transdermal delivery of narcotic analgesics: Comparative metabolism and permeability of human cadaver skin and hairless mouse skin, *J. Pharm. Sci.*, **83**, 1723–1728.

ROY, S.D., ROOS, E. and SHARMA, K., 1994b, Transdermal delivery of buprenorphine through cadaver skin, *J. Pharm. Sci.*, **83(2)**, 126–130.

ROY, S.D., CHATTERJEE, D.J., MANOUKIAN, E. and DIVOR, A., 1995a, Permeability of pure enantiomers of ketorolac through human cadaver skin, *J. Pharm. Sci.*, **84**, 987–990.

ROY, S.D., MANOUKIAN, E. and COMBS, D., 1995b, Absorption of transdermally delivered ketorolac acid in humans, *J. Pharm. Sci.*, **84**, 49–52.

ROY, S.D., GUTIERREZ, M., FLYNN, G.L. and CLEARY, G.W., 1996, Controlled transdermal delivery of fentanyl: Characterizations of pressure-sensitive adhesives for matrix patch design, *J. Pharm. Sci.*, **85**, 491–495.

ROYCHOWDHURY, S.M. and HEINRICHER, M.M., 1997, Effects of iontophoretically applied serotonin on three classes of physiologically characterized putative pain modulating neurons in the rostral ventromedial medulla of lightly anesthetized rats, *Neurosci. Lett.*, **226**, 136–138.

RUDDY, S.B. and HADZIJA, B.W., 1995, The role of stratum corneum in electrically facilitated transdermal drug delivery 1. Influence of hydration, tape-stripping and delipidization on the DC electrical properties of skin, *J. Control. Release*, **37**, 225–238.

RULAND, A., KREUTER, J. and RYTTING, J.H., 1994a, Transdermal delivery of the tetrapeptide hisetal (melanotropin (6–9)). I. Effect of various penetration enhancers: *In vitro* study across hairless mouse skin, *Int. J. Pharm.*, **101**, 57–61.

1994b, Transdermal delivery of the tetrapeptide hisetal (Melanotropin (6–9)): II. Effect of various penetration enhancers – *In vitro* study across human skin, *Int. J. Pharm.*, **103**, 77–80.

RUSSO, J., LIPMAN, A.G., COMSTOCK, T.J., PAGE, B.C. and STEPHEN, R.L., 1980, Lidocaine anesthesia: comparison of iontophoresis, injection and swabbing, *Am. J. Hosp. Pharm.*, **37**, 843–847.

SAGE, B.H., 1993, Iontophoresis, in: *Encyclopedia of Pharmaceutical Technology*. J. SWARBRICK and J.C. BOYLAN, (eds). Marcel Dekker, Inc., New York, pp. 217–247.

SAGE, B.H. and RIVIERE, J.E., 1992, Model systems in iontophoresis – Transport efficacy, *Adv. Drug Del. Rev.*, **9**, 265–287.
SAGE, B.H., BOCK, C.R., DENUZZIO, J.D. and HOKE, R.A., 1995, Technological and developmental issues of iontophoretic transport of peptide and protein drugs, in: *Trends and Future Perspectives in Peptide and Protein Drug Delivery.* V.H.L. LEE, M. HASHIDA, and Y. MIZUSHIMA, (eds). Harwood Academic Publishers GmbH, Chur, Switzerland, pp. 111–134.
SAGGINI, R., ZOPPI, M., VECCHIET, F., GATTESCHI, L., OBLETTER, G. and GIAMBERARDINO, M., 1996, Comparison of electromotive drug administration with ketorolac or with placebo in patients with pain from rheumatic disease: A double-masked study, *Clinical Ther.*, **18**, 1169–1174.
SANDERSON, J.E. and DERIEL, S.R., 1988, Methods and apparatus for iontophoretic drug delivery, U.S. Pat. 4722726.
SANDERSON, J.E., CALDWELL, R.W., HSIAO, J., DIXON, R. and TUTTLE, R.R., 1987, Noninvasive delivery of a novel inotropic catecholamine iontophoretic versus intravenous infusion in dogs, *J. Pharm. Sci.*, **76**, 215–218.
SANDERSON, J.E., RIEL, S.D. and DIXON, R., 1989, Iontophoretic delivery of nonpeptide drugs: Formulation optimization for maximum skin permeability, *J. Pharm. Sci.*, **78**, 361–364.
SANTI, P. and GUY, R.H., 1996a, Reverse iontophoresis – Parameters determining electroosmotic flow. II. Electrode chamber formulation, *J. Control. Release*, **42**, 29–36.
1996b, Reverse iontophoresis – Parameters determining electroosmotic flow. I. pH and ionic strength, *J. Control. Release*, **38**, 159–165.
SANTI, P., COLOMBO, P., BETTINI, R., CATELLANI, P.L., MINUTELLO, A. and VOLPATO, N.M., 1997, Drug reservoir composition and transport of salmon calcitonin in transdermal iontophoresis, *Pharm. Res.*, **14**, 63–66.
SANTOYO, S., ARELLANO, A., YGARTUA, P. and MARTIN, C., 1995, Penetration enhancer effects on the *in vitro* percutaneous absorption of piroxicam through rat skin, *Int. J. Pharm.*, **117**, 219–224.
SARPOTDAR, P.P., 1991, Iontophoresis, *Cosmet. Toilet.*, **106**, 94–100.
SARRAF, D. and LEE, D.A., 1994, The role of iontophoresis in ocular drug delivery, *J. Ocular Pharmacol.*, **10**, 69–81.
SATO, K., SUGIBAYASHI, K. and MORIMOTO, Y., 1991, Species differences in percutaneous absorption of nicorandil, *J. Pharm. Sci.*, **80**, 104–107.
SATYANAND, A., SAXENA, A.K. and AGARWAL, A., 1986, Silver iontophoresis in chronic osteomyelitis, *J. Indian Med. Assoc.*, **84**, 135–136.
SCANLON, K.J., OHTA, Y., ISHIDA, H., KIJIMA, H., OHKAWA, T., KAMINSKI, A., TSAI, J., HORNG, G. and KASHANISABET, M., 1995, Oligonucleotide-mediated modulation of mammalian gene expression, *FASEB J.*, **9**, 1288–1296.
SCHREIER, H. and BOUWSTRA, J., 1994, Liposomes and niosomes as topical drug carriers – Dermal and transdermal drug delivery, *J. Control. Release*, **30**, 1–15.
SCHWENDEMAN, S.P., AMIDON, G.L., LABHASETWAR, V. and LEVY, R.J., 1994, Modulated drug release using iontophoresis through heterogeneous cation-exchange membranes. 2. Influence of cation-exchanger content on membrane resistance and characteristic times, *J. Pharm. Sci.*, **83**, 1482–1494.
SCHWENDEMAN, S.P., LABHASETWAR, V. and LEVY, R.J., 1995, Model features of a cardiac iontophoretic drug delivery implant, *Pharm. Res.*, **12**, 790–795.
SCOTT, E.R., LAPLAZA, A.I., WHITE, H.S. and PHIPPS, J.B., 1993, Transport of ionic species in skin: Contribution of pores to the overall skin conductance, *Pharm. Res.*, **10**, 1699–1709.
SCOTT, E.R., PHIPPS, J.B. and WHITE, H.S., 1995, Direct imaging of molecular transport through skin, *J. Invest. Dermatol.*, **104**, 142–145.
SETH, S.C., ALLEN, L.V. and PINNAMARAJU, P., 1994, Stability of hydrocortisone salts during iontophoresis, *Int. J. Pharm.*, **106**, 7–14.
SHAH, A.K., WEI, G., LANMAN, R.C., BHARGAVA, V.O. and WEIR, S.J., 1996, Percutaneous absorption of ketoprofen from different anatomical sites in man, *Pharm. Res.*, **13**, 168–172.
SHAH, J.C., 1993, Analysis of permeation data – Evaluation of the lag time method, *Int. J. Pharm.*, **90**, 161–169.

SHAH, V.P., FLYNN, G.L., GUY, R.H., MAIBACH, H.I., SCHAEFER, H., SKELLY, J.P., WESTER, R.C. and YACOBI, A., 1991, In vivo percutaneous penetration/absorption, *Int. J. Pharm.*, **74**, 1–8.

SHAPIRO, B.L., PENCE, T.V. and WARWICK, W.J., 1975, Insulin iontophoresis in cystic fibrosis, *Proc. Soc. Exp. Biol. Med.*, **149**, 592–593.

SHARMA, V., STEBE, K., MURPHY, J.C. and TUNG, L., 1996, Poloxamer 188 decreases susceptibility of artificial lipid membranes to electroporation, *Biophys. J.*, **71**, 3229–3241.

SHAW, J.E., 1984, Pharmacokinetics of nitroglycerin and clonidine delivered by the transdermal route, *Am. Heart J.*, **108**, 217–223.

SHAW, K., 1997, The new transdermal technologies, *Pharmacy Times*, July, 38–41.

SHIN, B.C. and LEE, H.B., 1994, Iontophoretic delivery of insulin: Effects of skin treatments, *Proc. Int'l Symp. Control. Rel. Bioact. Mater.*, Controlled Release Society, Inc., 21, 373–374.

SHIVER, J.W., DAVIES, M.E., PERRY, H.C., FREED, D.C. and LIU, M.A., 1996a, Humoral and cellular immunities elicited by HIV-1 DNA vaccination, *J. Pharm. Sci.*, **85**, 1317–1324.

SHIVER, J.W., ULMER, J.B., DONNELLY, J.J. and LIU, M.A., 1996b, Humoral and cellular immunities elicited by DNA vaccines: Application to the human immunodeficiency virus and influenza, *Adv. Drug Del. Rev.*, **21**, 19–31.

SHRIVASTAVA, S.N. and SINGH, G., 1977, Tap water iontophoresis in palmoplantar hyperhidrosis, *Br. J. Dermatol.*, **96**, 189–195.

SIDDIQUI, O., ROBERTS, M.S. and POLACK, A.E., 1985, The effect of iontophoresis and vehicle pH on the *in vitro* permeation of lignocaine through human stratum corneum, *J. Pharm. Pharmacol.*, **37**, 732–735.

SIDDIQUI, O., SUN, Y., LIU, J.C. and CHIEN, Y.W., 1987, Facilitated transdermal transport of insulin, *J. Pharm. Sci.*, **76**, 341–345.

SIMMONDS, M.A., 1995, Transdermal fentanyl: Clinical development in the United States, *Anti-Cancer Drug*, **6**, 35–38.

SIMONIN, J.P., 1995, On the mechanisms of *in vitro* and *in vivo* phonophoresis, *J. Control. Release*, **33**, 125–141.

SIMS, S.M., HIGUCHI, W.I. and SRINIVASAN, V., 1991, Interaction of electric field and electroosmotic effects in determining iontophoretic enhancement of anions and cations, *Int. J. Pharm.*, **77**, 107–118.

1992, Skin alteration and convective solvent flow effects during iontophoresis II. Monovalent anion and cation transport across human skin, *Pharm. Res.*, **9**, 1402–1409.

SINGH, J. and BHATIA, K.S., 1996, Topical iontophoretic drug delivery: Pathways, principles, factors, and skin irritation, *Med. Res. Rev.*, **16**, 285–296.

SINGH, J. and ROBERTS, M.S., 1989, Transdermal delivery of drugs by iontophoresis: A review, *Drug Design Del.*, **4**, 1–12.

SINGH, J., GROSS, M., O'CONNELL, M., SAGE, B., and MAIBACH, H.I., 1994, Effect of iontophoresis in different ethnic groups' skin function, *Proc. Int'l. Symp. Control. Rel. Bioact. Mater.*, Controlled Release Society, Inc., 21. pp. 365–366.

SINGH, P. and MAIBACH, H.I., 1994a, Iontophoresis in drug delivery: Basic principles and applications, *Crit. Rev. Ther. Drug Carr. Syst.*, **11**, 161–213.

1994b, Transdermal iontophoresis: Pharmacokinetic considerations, *Clin. Pharmacokinet.*, **26**(5), 327–334.

1996, Iontophoresis: An alternative to the use of carriers in cutaneous drug delivery, *Adv. Drug Del. Rev.*, **18**, 379–394.

SINGH, P. and ROBERTS, M.S., 1993, Iontophoretic transdermal delivery of salicyclic acid and lidocaine to local subcutaneous structures, *J. Pharm. Sci.*, **82**, 127–131.

SINGH, P., ANLIKER, M., SMITH, G.A., ZAVORTINK, D. and MAIBACH, H.I., 1995, Transdermal iontophoresis and solute penetration across excised human skin, *J. Pharm. Sci.*, **84**, 1342–1346.

SINGH, P., BONIELLO, S., LIU, P. and DINH, S., 1997, Iontophoretic transdermal delivery of methylphenidate hydrochloride, *Pharm. Res.*, **14**, S309.

SJOBERG, H., KARAMI, K., BERONIUS, P. and SUNDELOF, L.O., 1996, Ionization conditions for iontophoretic drug delivery. A revised pK(a) of lidocaine hydrochloride in aqueous solution at 25 degrees C established by precision conductometry, *Int. J. Pharm.*, **141**, 63–70.

SKYDSGAARD, M. and HOUNSGAARD, J., 1996, Multiple actions of iontophoretically applied serotonin on motorneurones in the turtle spinal cord *in vitro*, *Acta Physiol. Scand.*, **158**, 301–310.

SLAVIKOVA, K. and MASSOURIDOU, E., 1995, DNA-mediated gene transfer into mammalian cells and cancer, *Neoplasma.*, **42**, 293–297.

SLOAN, J.B. and SOLTANI, K., 1986, Iontophoresis in dermatology, *J. Am. Acad. Dermatol.*, **15**, 671–684.

SLOAN, K.B., BEALL, H.D., WEIMAR, W.R. and VILLANUEVA, R., 1991, The effect of receptor phase composition on the permeability of hairless mouse skin in diffusion cell experiments, *Int. J. Pharm.*, **73**, 97–104.

SLOUGH, C.L., SPINELLI, M.J. and KASTING, G.B., 1988, Transdermal delivery of etidronate (EHDP) in the pig via iontophoresis, *J. Memb. Sci.*, **35**, 161–165.

SMITH, K.J., KONZELMAN, J.L., LOMBARDO, F.A., SKELTON, H.G., HOLLAND, T.T., YEAGER, J., WAGNER, K.F., OSTER, C.N. and CHUNG, R., 1992, Iontophoresis of vinblastine into normal skin and for treatment of Kaposi's sarcoma in human immunodeficiency virus-positive patients, *Arch. Dermatol.*, **128**, 1365–1370.

SOPHIE, T. and VERONIQUE, P., 1991, Iontophoretic mobility of insulin across hairless rat skin, *Proc. Int'l Symp. Control. Rel. Bioact. Mater.*, Controlled Release Society, Inc., 18, pp. 609–610.

SOUTHAM, M.A., 1995, Transdermal fentanyl therapy: System design, pharmacokinetics and efficacy, *Anti-Cancer Drug*, **6**, 29–34.

SRINIVASAN, V., HIGUCHI, W.I., SIMS, S.M., GHANEM, A.H. and BEHL, C.R., 1989, Transdermal iontophoretic drug delivery: Mechanistic analysis and application to polypeptide delivery, *J. Pharm. Sci.*, **78**, 370–375.

SRINIVASAN, V., SU, M., HIGUCHI, W.I. and BEHL, C.R., 1990, Iontophoresis of polypeptides: Effect of ethanol pretreatment of human skin, *J. Pharm. Sci.*, **79**, 588–591.

STAGNI, G., O'DONNELL, D., LIU, Y.J., KELLOGG, D.L. and SHEPHERD, A.M.M., 1997a, Evaluation of the interaction between iontophoretic current and intradermal microdialysis in human forearm skin, *Pharm. Res.*, **14**, S70.

1997b, Evaluation of an intradermal microdialysis technique to study transdermal iontophoretic drug delivery in the forearm of human subjects: Preliminary results, *Pharm. Res.*, **14**, S69.

STEINSTRASSER, I. and MERKLE, H.P., 1995, Dermal metabolism of topically applied drugs: Pathways and models reconsidered, *Pharm. Acta Helv.*, **70**, 3–24.

STEPHEN, R.L., PETELENZ, T.J. and JACOBSEN, S.C., 1984, Potential novel methods for insulin administration: I. Iontophoresis, *Biomed. Biochim. Acta*, **43**, 553–558.

1990, Method of iontophoretically treating acne, furuncles and like skin disorders, U.S. Pat. 4979938. Assignee: Iomed.

STEPHEN, R.L., BONEZZI, C., ROSSI, C. and ERUZZI, S., 1997, Morphine formulations for use by electromotive administration, U.S. Pat. 5607940.

STEPHENS, W.G.S., 1963, The current–voltage relationship in human skin, *Med. Electron. Biol. Eng.*, **1**, 389–399.

STEVENS, M.F., LINSTEDT, U., NERUDA, B., LIPFERT, P. and WULF, H., 1996, Effect of transcutaneous electrical nerve stimulation on onset of axillary plexus block, *Anaesthesia*, **51**, 916–919.

STEWART, F.M., 1995, Getting human gene therapy to work, *J. Cell Biochem.*, **58**, 416–423.

STOLMAN, L.P. 1987, Treatment of excess sweating of the palms by iontophoresis, *Arch. Dermatol.*, **123**, 893–896.

STONE, T.W., 1972, Responses of blood vessels to various amines applied by microiontophoresis, *J. Pharm. Pharmacol.*, **24**, 318–323.

STORM, G., KOPPENHAGEN, F., HEEREMANS, A., VINGERHOEDS, M., WOODLE, M.C. and CROMMELIN, D.J.A., 1995, Novel developments in liposomal delivery of peptides and proteins, *J. Control. Release*, **36**, 19–24.

STRALKA, S.W., HEAD, P.L. and MOHR, K., 1996, The clinical use of iontophoresis, *Phys. Ther. Prod.*, March, 48–51.

STRATTAN, C.E., 1992a, 2-Hydroxypropyl-beta-cyclodextrin: Part I, Patents and regulatory issues, *Pharm. Technol.*, **16**(1), 69–74.

1992b, 2-Hydroxypropyl-beta-cyclodextrin: Part II, Safety and manufacturing issues, *Pharm. Technol.*, **16(2)**, 52–58.
SU, M.H., SRINIVASAN, V., GHANEM, A.H. and HIGUCHI, W.I., 1994, Quantitative *in vivo* iontophoretic studies, *J. Pharm. Sci.*, **83**, 12–17.
SUN, Y. and XUE, H., 1991, Transdermal delivery of insulin by iontophoresis, *Ann. N. Y. Acad. Sci.*, **618**, 596–598.
SWARBRICK, J., LEE, G., BROM, J. and GENSMANTEL, P., 1984, Drug permeation through human skin II: permeability of ionizable compounds, *J. Pharm. Sci.*, **73**, 1352–1355.
SZEJTLI, J. 1991a, Cyclodextrins in drug formulations: Part II, *Pharm. Technol.*, **15**, 24–38.
1991b, Cyclodextrins in drug formulations: Part I, *Pharm. Technol.*, **15**, 36–44.
1994, Medicinal applications of cyclodextrins, *Med. Res. Rev.*, **14**, 353–386.
TACHIBANA, K. 1992, Transdermal delivery of insulin to alloxan-diabetic rabbits by ultrasound exposure, *Pharm. Res.*, **9**, 952–954.
TACHIBANA, K. and TACHIBANA, S., 1991, Transdermal delivery of insulin by ultrasonic vibration, *J. Pharm. Pharmacol.*, **43**, 270–271.
TANNENBAUM, M., 1980, Iodine iontophoresis in reducing scar tissue, *Phys. Ther.*, **60**, 792.
TAPPER, R., 1993, Iontophoretic treatment system, U.S. Pat. 5224927. Assignee: Tapper R.
1997, Iontophoretic delivery of medicaments through skin – with electrical treatment current between electrodes periodically reversed at very low frequencies to mitigate tissue damage, Pat. EP 776676. Assignee: Tapper R.
TASCON, R.E., COLSTON, M.J., RAGNO, S., STAVROPOULOS, E., GREGORY, D. and LOWRIE, D.B., 1996, Vaccination against tuberculosis by DNA injection, *Nature Med.*, **2**, 888–892.
TATSUKA, M., YAMAGISHI, N., WADA, M., MITSUI, H., OTA, T. and ODASHIMA, S., 1995, Electroporation-mediated transfection of mammalian cells with crude plasmid DNA preparations, *Genet. Anal. Biomol. Eng.*, **12**, 113–117.
TAUBER, U., 1989, Drug metabolism in the skin: Advantages and disadvantages, in: *Transdermal Drug Delivery: Developmental issues and research initiatives*, J. HADGRAFT and R.H. GUY, (eds). Marcel Dekker Inc., New York pp. 99–134.
1994, Dermatocorticosteroids: Structure, activity, pharmacokinetics, *Eur. J. Dermatol.*, **4**, 419–429.
TERZO, S.D., BEHL, C.R. and NASH, R.A., 1989, Iontophoretic transport of a homologous series of ionized and nonionized model compounds: Influence of hydrophobicity and mechanistic interpretation, *Pharm. Res.*, **6**, 85–90.
TESSARI, L., CECILIANI, L., BELLUATI, A., LETIZIA, G., MARTORANA, U., PAGLIARA, L., POGNANI, A., THOVEZ, G., SICLARI, A., TORRI, G., SOLIMENO, L. and MONTULL, E., 1995, Aceclofenac cream versus piroxicam cream in the treatment of patients with minor traumas and phlogistic affections of soft tissues: A double-blind study, *Curr. Ther. Res.*, **56**, 702–712.
THACHARODI, D. and RAO, K.P., 1996, Collagen membrane controlled transdermal delivery of propranolol hydrochloride, *Int. J. Pharm.*, **131**, 97–99.
THEEUWES, F., GYORY, J.R. and HAAK, R.P., 1992, Membrane for electrotransport transdermal drug delivery, U.S. Pat. 5080646. Assignee: Alza Corp.
THYSMAN, S. and PREAT, V., 1993, *In vivo* iontophoresis of fentanyl and sufentanil in rats: Pharmacokinetics and acute antinociceptive effects, *Anesth. Analg.*, **77**, 61–66.
THYSMAN, S., PREAT, V. and ROLAND, M., 1992, Factors affecting iontophoretic mobility of metoprolol, *J. Pharm. Sci.*, **81**, 670–675.
THYSMAN, S., HANCHARD, C. and PREAT, V., 1994a, Human calcitonin delivery in rats by iontophoresis, *J. Pharm. Pharmacol.*, **46**, 725–730.
THYSMAN, S., TASSET, C. and PREAT, V., 1994b, Transdermal iontophoresis of fentanyl – Delivery and mechanistic analysis, *Int. J. Pharm.*, **101**, 105–113.
THYSMAN, S., JADOUL, A., LEROY, T., VANNESTE, D. and PREAT, V., 1995a, Laser doppler evaluation of skin reaction in volunteers after histamine iontophoresis, *J. Control. Release*, **36**, 215–219.
THYSMAN, S., VANNESTE, D. and PREAT, V., 1995b, Noninvasive investigation of human skin after *in vivo* iontophoresis, *Skin Pharmacol.*, **8**, 229–236.

TITOMIROV, A.V., SUKHAREV, S. and KISTANOVA, E., 1991, In vivo electroporation and stable transformation of skin cells of newborn mice by plasmid DNA, *Biochim. Biophys. Acta*, **1088**, 131–134.
TOJO, K., 1987, Random brick model for drug transport across stratum corneum, *J. Pharm. Sci.*, **76**, 889–891.
TOJO, K., SUN, Y., GHANNAM, M.M. and CHIEN, Y.W., 1985, Characterization of a membrane permeation system for controlled drug delivery studies, *AICHE J.*, **31**, 741–746.
TOMER, R., DIMITRIJEVIC, D. and FLORENCE, A.T., 1995, Electrically controlled release of macromolecules from cross-linked hyaluronic acid hydrogels, *J. Control. Release*, **33**, 405–413.
TOMOV, T.C., 1995, Quantitative dependence of electroporation on the pulse parameters, *Bioelectrochem. Bioenerg.*, **37**, 101–107.
TORRY, M., WILCOCK, A., COOPER, B.G. and TATTERSFIELD, A.E., 1996, The effect of chest wall transcutaneous electrical nerve stimulation on dyspnoea, *Resp. Physiol.*, **104**, 23–28.
TOUITOU, E., SHACO-EZRA, N., DAYAN, N., JUSHYNSKI, M., RAFAELOFF, R. and AZOURY, R., 1992, Dyphylline liposomes for delivery to the skin, *J. Pharm. Sci.*, **81**, 131–134.
TRAINER, A.H. and ALEXANDER, M.Y., 1997, Gene delivery to the epidermis, *Hum. Mol. Genet.*, **6**, 1761–1767.
TRANSDERMAL NICOTINE STUDY GROUP, 1991, Transdermal nicotine for smoking cessation: Six-month results from two multicenter controlled clinical trials, *JAMA*, **266**, 3133–3138.
TRECO, D.A. and SELDEN, R.F., 1995, Non-viral gene therapy, *Mol. Med. Today*, **1**, 314–321.
TSCHAN, T., STEFFEN, H. and SUPERSAXO, A., 1997, Sebaceous-gland deposition of isotretinoin after topical application: An in vitro study using human facial skin, *Skin Pharmacol.*, **10**, 126–134.
TSONG, T.Y., 1991, Electroporation of cell membranes, *Biophys. J.*, **60**, 297–306.
TU, Y.H. and ALLEN, L.V., 1989, In vitro iontophoretic studies using synthetic membranes, *J. Pharm. Sci.*, **78**, 211–213.
TURNER, N.G., KALIA, Y.N. and GUY, R.H., 1997, The effect of current on skin barrier function in vivo: Recovery kinetics post-iontophoresis, *Pharm. Res.*, **14**, 1252–1257.
TYLE, P. and AGRAWALA, P. 1989, Drug delivery by phonophoresis, *Pharm. Res.*, **6**, 355–361.
UCHIDA, S., MORISHITA, T., IKEDA, Y. and AKASHI, T., 1995, Anti-inflammatory effect of flurbiprofen tape applied percutaneously to rats with adjuvant-induced arthritis, *Jpn. J. Pharmacol.*, **69**, 37–41.
UEDA, H., OGIHARA, M., SUGIBAYASHI, K. and MORIMOTO, Y., 1996, Change in the electrochemical properties of skin and the lipid packing in stratum corneum by ultrasonic irradiation, *Int. J. Pharm.*, **137**, 217–224.
UNTEREKER, D.F., PHIPPS, J.B. and LATTIN, G.A., 1996, Iontophoretic drug delivery, U.S. Pat. 5573503. Assignee: Alza Corp.
VALIA, K.H., TOJO, K. and CHIEN, Y.W., 1985, Long-term permeation kinetics of estradiol: (III) Kinetic analysis of the simultaneous skin permeation and bioconversion of estradiol esters, *Drug Dev. Ind. Pharm.*, **11**, 1133–1173.
VALLE, C.A.L., GERMAIN, L., ROUABHIA, M., XU, W., GUIGNARD, R., GOULET, F. and AUGER, F.A., 1996, Grafting on nude mice of living skin equivalents produced using human collagens, *Transplantation*, **62**, 317–323.
VANBEVER, R., LECOUTURIER, N. and PREAT, V., 1994, Transdermal delivery of metoprolol by electroporation, *Pharm. Res.*, **11**, 1657–1662.
VANBEVER, R., LE BOULENGE, E. and PREAT, V., 1996, Transdermal delivery of fentanyl by electroporation. 1. Influence of electrical factors, *Pharm. Res.*, **13**, 559–565.
VANBEVER, R., PRAUSNITZ, M.R. and PREAT, V., 1997, Macromolecules as novel transdermal transport enhancers for skin electroporation, *Pharm. Res.*, **14**, 638–644.
VAN BUSKIRK, G.A., GONZALEZ, M.A., SHAH, W.P., BARNHARDT, S., BARRETT, C., BERGE, S., CLEARY, G., CHAN, K., FLYNN, G., FOSTER, T., GALE, R., GARRISON, R., GOCHNOUR, S., GOTTO, A., GOVIL, S., GRAY, V.A., HAMMAR, J., HARDER, S., HOIBERG, C., HUSSAIN, A., KARP, C., LLANOS, H., MANTELLE, J., NOONAN, P., SWANSON, D. and ZERBE, H., 1997, Scale-up of adhesive transdermal drug delivery systems, *Pharm. Res.*, **14**, 848–852.

VAN DER GEEST, R., HUEBER, F., SZOKA, F.C. and GUY, R.H., 1996a, Iontophoresis of bases, nucleosides, and nucleotides, *Pharm. Res.*, **13**, 553–558.

VAN DER GEEST, R., ELSHOVE, D.A.R., DANHOF, M., LAVRIJSEN, A.P.M. and BODDE, H.E., 1996b, Non-invasive assessment of skin barrier integrity and skin irritation following iontophoretic current application in humans, *J. Control. Release*, **41**, 205–213.

VAN DER MOLEN, R.G., SPIES, F., VANT NOORDENDE, J.M., BOELSMA, E., MOMMAAS, A.M. and KOERTEN, H.K., 1997, Tape stripping of human stratum corneum yields cell layers that originate from various depths because of furrows in the skin, *Arch. Dermatol. Res.*, **289**, 514–518.

VAN HAL, D.A., JEREMIASSE, E., JUNGINGER, H.E., SPIES, F. and BOUWSTRA, J.A., 1996, Structure of fully hydrated human stratum corneum: A freeze fracture electron microscopy study, *J. Invest. Dermatol.*, **106**, 89–95.

VAN NESTE, D., LEROY, D.T., DEBROUWER, B. and RIHOUX, J.P., 1996, Histamine-induced skin reactions using iontophoresis and H_1- blockade, *Inflamm. Research.*, **45**, S48–S49.

VECCHINI, L. and GROSSI, E., 1984, Ionization with diclofenac sodium in rheumatic disorders: A double-blind placebo-controlled trial, *J. Int. Med. Res.*, **12**, 346–350.

VICKERY, B.H., 1991, Biological actions of synthetic analogues of luteinizing hormone-releasing hormone, in: *Pharmacokinetics and Pharmacodynamics, Vol.3: Peptides, Peptoids and Proteins*, P.D. GARZONE, W.A. COLBURN, and M. MOKOTOFF, (eds). Harvey Whitney Books Co., Cincinnati, Ohio, pp. 41–49.

VITORIA, M., BENTLEY, L.B., VIANNA, R.F., WILSON, S. and COLLETT, J.H. 1997, Characterization of the influence of some cyclodextrins on the stratum corneum from the hairless mouse, *J. Pharm. Pharmacol.*, **49**, 397–402.

VOLLMER, U. and CORDES, G., 1997, Physical quality control of transdermal delivery systems, *Proc. Int'l Symp. Control. Rel. Bioact. Mater.*, Controlled Release Society, Inc., 24, pp. 685–686.

VOLLMER, U., MULLER, B.W., WILFFERT, B. and PETERS, T., 1993, An improved model for studies on transdermal drug absorption *in vivo* in rats, *J. Pharm. Pharmacol.*, **45**, 242–245.

VOLLMER, U., MULLER, B.W., PEETERS, J., MESENS, J., WILFFERT, B. and PETERS, T., 1994, A study of the percutaneous absorption-enhancing effects of cyclodextrin derivatives in rats, *J. Pharm. Pharmacol.*, **46**, 19–22.

VOLPATO, N.M., SANTI, P. and COLOMBO, P., 1995, Iontophoresis enhances the transport of acyclovir through nude mouse skin by electrorepulsion and electro-osmosis, *Pharm. Res.*, **12**, 1623–1627.

VUTLA, N.B., BETAGERI, G.V. and BANGA, A.K., 1996, Transdermal iontophoretic delivery of enkephalin formulated in liposomes, *J. Pharm. Sci.*, **85**, 5–8.

WAHLBERG, J.E., 1970, Skin clearance of iontophoretically administered chromium (^{51}Cr) and sodium (^{22}Na) ions in the guinea pig, *Acta Derm. Venereol. (Oslo)*, **50**, 255–262.

WALKER, R.B. and SMITH, E.W., 1996, The role of percutaneous penetration enhancers, *Adv. Drug Del. Rev.*, **18**, 295–301.

WALTERS, K.A. and ROBERTS, M.S., 1993, Veterinary applications of skin penetration enhancers, in: *Pharmaceutical Skin Penetration Enhancement*. K.A. WALTERS and J. HADGRAFT, (eds). Marcel Dekker, Inc., New York, pp. 345–364.

WANG, B., BOYER, J., SRIKANTAN, V., UGEN, K., GILBERT, L., PHAN, C., DANG, K., MERVA, M., AGADJANYAN, M.G., NEWMAN, M., CARRANO, R., MCCALLUS, D., CONEY, L., WILLIAMS, W.V. and WEINER, D.B. 1995, Induction of humoral and cellular immune responses to the human immunodeficiency type 1 virus in nonhuman primates by *in vivo* DNA inoculation, *Virology*, **211**, 102–112.

WANG, S., KARA, M. and KRISHNAN, T.R., 1997, Topical delivery of cyclosporin A coevaporate using electroporation technique, *Drug Dev. Ind. Pharm.*, **23**, 657–663.

WANG, Y., ALLEN, L.V., LI, L.C. and TU, Y.H., 1993, Iontophoresis of hydrocortisone across hairless mouse skin: investigation of skin alteration, *J. Pharm. Sci.*, **82**, 1140–1144.

WARDS, B.J. and COLLINS, D.M., 1996, Electroporation at elevated temperatures substantially improves transformation efficiency of slow-growing mycobacteria, *FEMS Microbiol. Lett.*, **145**, 101–105.

WARREN, R.A. and DYKES, R.W., 1996, Transient and long-lasting effects of iontophoretically administered norepinephrine on somatosensory cortical neurons in halothane-anesthetized cats, *Can. J. Physiol. Pharmacol.*, **74**, 38–57.
WARWICK, W.J., HUANG, N.N., WARING, W.W., CHERIAN, A.G., BROWN, I., STEJSKAL-LORENZ, E., YEUNG, W.H., DUHON, G., HILL, J.G. and STROMINGER, D., 1986, Evaluation of a cystic fibrosis screening system incorporating a miniature sweat stimulator and disposable chloride sensor, *Clin. Chem.*, **32**, 850–853.
WARWICK, W.J., HANSEN, L.G. and WERNESS, M.E., 1990, Quantitation of chloride in sweat with the cystic fibrosis indicator system, *Clin. Chem.*, **36**, 96–98.
WAUD, D.R., 1967, Iontophoretic application of drugs, *J. Appl. Physiol.*, **23**, 128–130.
WEARLEY, L., LIU, J.C. and CHIEN, Y.W., 1989a, Iontophoresis facilitated transdermal delivery of verapamil. I. *In vitro* evaluation and mechanistic studies, *J. Control. Release*, **8**, 237–250.
1989b, Iontophoresis facilitated transdermal delivery of verapamil. II. Factors affecting the reversibility of skin permeability, *J. Control. Release*, **9**, 231–242.
WEARLEY, L.L., TOJO, K. and CHIEN, Y.W., 1990, A numerical approach to study the effect of binding on the iontophoretic transport of a series of amino acids, *J. Pharm. Sci.*, **79**, 992–998.
WEAVER, J.C., 1993, Electroporation: A general phenomenon for manipulating cells and tissues, *J. Cellular Biochem.*, **51**, 426–435.
WEAVER, J.C., 1995, Electroporation theory – Concepts and mechanisms, in: *Electroporation Protocols for Microorganisms*, J.A. NICKOLOFF (ed.). Humana Press, Totowa, NJ, pp. 1–26.
WEAVER, J.C. and CHIZMADZHEV, Y., 1996, Electroporation, in: *Biological Effects of Electromagnetic Fields*, C. POLTE and E. POSTOW (eds), 2nd edn. CRG Press, Inc., Boca Raton, NY, pp. 247–274.
WEAVER, J.C., POWELL, K.T. and LANGER, R.S., 1991, Control of transport of molecules across tissue using electroporation, U.S. Pat. 5 019 034. Assignee: MIT.
WEAVER, J.C., VANBEVER, R., VAUGHAN, T.E. and PRAUSNITZ, M.R., 1997, Heparin alters transdermal transport associated with electroporation, *Biochem. Biophys. Res. Commun.*, **234**, 637–640.
WEBSTER, H.L., 1983, Iontophoretic electrode device, method and gel insert, U.S. Pat. 4383529. Assignee: Wescor.
WEBSTER, H.L. and BARLOW, W.K., 1985, Sweat-collecting device and method, U.S. Pat. 4542751. Assignee: Wescor.
WEINER, N., WILLIAMS, N., BIRCH, G., RAMACHANDRAN, C., SHIPMAN, C. and FLYNN, G., 1989, Topical delivery of liposomally encapsulated interferon evaluated in a cutaneous herpes guinea pig model, *Antimicrob. Agents Chemother.*, **33**, 1217–1221.
WEINER, N., LIEB, L., NIEMIEC, S., RAMACHANDRAN, C., HU, Z. and EGBARIA, K., 1994, Liposomes: A novel topical delivery system for pharmaceutical and cosmetic applications, *J. Drug Target.*, **2**, 405–410.
WEINSHENKER, N.M. and O'NEILL, W.P., 1991, Pharmaceutical preparations containing cyclodextrins and their use in iontophoretic therapies. U.S. Pat. 5068226. Assignee: Cyclex.
WERTZ, P.W., ABRAHAM, W., LANDMANN, L. and DOWNING, D.T., 1986, Preparation of liposomes from stratum corneum lipids, *J. Invest. Dermatol.*, **87**, 582–584.
WESTER, R.C. and MAIBACH, H.I., 1992, Percutaneous absorption of drugs, *Clin. Pharmacokinet.*, **23**, 253–266.
WHEELER, C.J., FELGNER, P.L., TSAI, Y.J., MARSHALL, J., SUKHU, L., DOH, S.G., HARTIKKA, J., NIETUPSKI, J., MANTHORPE, M., NICHOLS, M., PLEWE, M., LIANG, X.W., NORMAN, J., SMITH, A. and CHENG, S.H., 1996, A novel cationic lipid greatly enhances plasmid DNA delivery and expression in mouse lung, *Proc. Natl. Acad. Sci. USA*, **93**, 11454–11459.
WICKS, I., 1995, Human gene therapy, *Aust. NZ. J. Med.*, **25**, 280–283.
WIEDER, D.L., 1992, Treatment of traumatic myositis ossificans with acetic acid iontophoresis, *Phys. Ther.*, **72**, 133–137.
WILDING, I.R., DAVIS, S.S., RIMOY, G.H., RUBIN, P., KURIHARABERGSTROM, T., TIPNIS, V., BERNER, B. and NIGHTINGALE, J., 1996, Pharmacokinetic evaluation of transdermal buprenorphine in man, *Int. J. Pharm.*, **132**, 81–87.

WILLIAMS, P.L. and RIVIERE, J.E., 1993, Model describing transdermal iontophoretic delivery of lidocaine incorporating consideration of cutaneous microvascular state, *J. Pharm. Sci.*, **82**, 1080–1084.

1995, A biophysically based dermatopharmacokinetic compartment model for quantifying percutaneous penetration and absorption of topically applied agents. 1. Theory, *J. Pharm. Sci.*, **84**, 599–608.

WINDISCH, V., DE LUCCIA, F., DUHAU, L., HERMAN, F., MENCEL, J.J., TANG, S.Y. and VUILHORGNE, M., 1997, Degradation pathways of salmon calcitonin in aqueous solution, *J. Pharm. Sci.*, **86**, 359–364.

WINTZ, H. and DIETRICH, A., 1996, Electroporation of small RNAs into plant protoplasts: Mitochondrial uptake of transfer RNAs, *Biochem. Biophys. Res. Commun.*, **223**, 204–210.

WOODSON, L.P., 1995, Biofilm reduction method, U.S. Pat. 5462644. Assignee: MN Mining & Mfg.

WU, P.C., HUANG, Y.B., LIN, H.H. and TSAI, Y.H., 1996, Percutaneous absorption of captropil from hydrophilic cellulose gel(R) through excised rabbit skin and human skin, *Int. J. Pharm.*, **145**, 215–220.

WU, P.C., FANG, J.Y., HUANG, Y.B. and TSAI, Y.H., 1997, Development and evaluation of transdermal patches of nonivamide and sodium nonivamide acetate, *Pharmazie.*, **52**, 135–138.

XIONG, G.L., QUAN, D.Y. and MAIBACH, H.I., 1996, Effects of penetration enhancers on in vitro percutaneous absorption of low molecular weight heparin through human skin, *J. Control. Release*, **42**, 289–296.

YAMAMOTO, T. and YAMAMOTO, Y., 1976, Electrical properties of the epidermal stratum corneum, *Med. Biol. Eng.*, **14**, 151–158.

YAMASHITA, F., KOYAMA, Y., SEZAKI, H. and HASHIDA, M., 1993, Estimation of a concentration profile of acyclovir in the skin after topical administration, *Int. J. Pharm.*, **89**, 199–206.

YANG, N., BURKHOLDER, J., ROBERTS, B., MARTINELL, B. and McCABE, D., 1990, In vivo and in vitro gene transfer to mammalian somatic cells by particle bombardment, *Proc. Natl. Acad. Sci. USA*, **87**, 9568–9572.

YANG, N.S. and SUN, W.H., 1995, Gene gun and other non-viral approaches for cancer gene therapy, *Nature Med.*, **1**, 481–483.

YANG, S.H., SONG, T.S. and KIM, Y.M., 1996, Transformation of methylotrophic bacteria by electroporation of pRO1727 plasmid from *Pseudomonas aeruginosa*, *Mol. Cells*, **6**, 225–228.

YEUNG, W.H., PALMER, J., SCHIDLOW, D., BYE, M.R. and HUANG, N.N., 1984, Evaluation of a paper-patch test for sweat chloride determination, *Clin. Pediatrics*, **23**, 603–607.

YOSHIDA, N.H. and ROBERTS, M.S., 1992, Structure–transport relationships in transdermal iontophoresis, *Adv. Drug Del. Rev.*, **9**, 239–264.

1993, Solute molecular size and transdermal iontophoresis across excised human skin, *J. Control. Release*, **25**, 177–195.

1994, Role of conductivity in iontophoresis. 2. Anodal iontophoretic transport of phenylethylamine and sodium across excised human skin, *J. Pharm. Sci.*, **83**, 344–350.

1995, Prediction of cathodal iontophoretic transport of various anions across excised skin from different vehicles using conductivity measurements, *J. Pharm. Pharmacol.*, **47**, 883–890.

YU, H.Y. and LIAO, H.M., 1996, Triamcinolone permeation from different liposome formulations through rat skin *in vitro*, *Int. J. Pharm.*, **127**, 1–7.

ZAKZEWSKI, C.A. and LI, J.K., 1991, Pulsed mode constant current iontophoretic transdermal metoprolol tartrate delivery in established acute hypertensive rabbits, *J. Control. Release*, **17**, 157–162.

ZAKZEWSKI, C.A., LI, J.K.J., AMORY, D.W., JENSEN, J.C. and KALATZIS MANOLAKIS, E., 1996, Design and implementation of a constant-current pulsed iontophoretic stimulation device, *Med. Biol. Eng. Comput.*, **34**, 484–488.

ZANKEL, H.T. and DURHAM, N.C., 1963, Effect of physical modalities upon Ra I^{131} iontophoresis, *Arch. Phys. Med. Rehabil.*, **44**, 93–97.

ZANKEL, H.T., CRESS, R.H. and KAMIN, H., 1959, Iontophoresis studies with a radioactive tracer, *Arch. Phys. Med. Rehabil.*, **40**, 193–196.

ZELPHATI, O. and SZOKA, F.C., 1996, Liposomes as a carrier for intracellular delivery of antisense oligonucleotides: A real or magic bullet? *J. Control. Release*, **41**, 99–119.

ZELTZER, L., REGALADO, M., NICHTER, L.S., BARTON, D., JENNINGS, S. and PITT, L., 1991, Iontophoresis versus subcutaneous injection: A comparison of two methods of local anesthesia delivery in children, *Pain*, **44**, 73–78.

ZEWERT, T.E., PLIQUETT, U.F., LANGER, R. and WEAVER, J.C., 1995, Transdermal transport of DNA antisense oligonucleotides by electroporation, *Biochem. Biophys. Res. Commun.*, **212**, 286–292.

ZHANG, L. and HOFMANN, G.A., 1997, Electric pulse mediated transdermal drug delivery, *Proc. Int'l Symp. Control. Rel. Bioact. Mater.*, Controlled Release Society, Inc., 24, pp. 27–28.

ZHANG, L., LI, L.N., HOFFMANN, G.A. and HOFFMAN, R.M., 1996, Depth-targeted efficient gene delivery and expression in the skin by pulsed electric fields: An approach to gene therapy of skin aging and other diseases, *Biochem. Biophys. Res. Commun.*, **220**, 633–636.

ZHANG, L., LI, L.N., AN, Z.L., HOFFMAN, R.M. and HOFMANN, G.A., 1997, *In vivo* transdermal delivery of large molecules by pressure-mediated electroincorporation and electroporation: a novel method for drug and gene delivery, *Bioelectrochem. Bioenerg.*, **42**, 283–292.

ZIERHUT, M., BIEBER, T., BROCKER, E.B., FORRESTER, J.V., FOSTER, C.S. and STREILEIN, J.W., 1996, Immunology of the skin and the eye, *Immunol. Today*, **17**, 448–450.

Index

acyclovir 22
adverse effects
   electroporation 103–5
   iontophoresis 126–30
Alza Corporation 119–21
amino acids 21, 77, 82–3
angiotensin 19
antibiotics 66–7
antiviral agents 66
arthritis, treatment of 58

battery technology 132
becton dickinson 121–2
beta blockers 25, 28–9
bioavailability 125–6
buffer systems 22–3
buprenorphine 30

calcein 101
calcitonin 35, 88–9
capillary electrophoresis 79
captopril 29
carpal tunnel syndrome, treatment of 58
chemical enhancers 11–12, 81–2
chemotherapy by electroporation 106–7
colchicine 51
commercial development
   electroporation 107–8
   iontophoresis 119–23, 132–4
current
   density 19
   pulsed 20–1
cyclodextrins 55
Cygnus, Inc. 122
cystic fibrosis, diagnosis of 63–5

developmental issues 132–4
dexamethasone 21, 57–9, 74
diagnostic applications 26–8
diffusion cells
   horizontal configuration 39
   vertical configuration 39–40

DNA vaccines 115–17
dose delivered 62, 69, 125
Drionic® unit 63, 69
drug
   analysis 42–3
   conductivity 21
   formulation 22
   physicochemical properties 21–2
   size limit 22
DUPEL® device 69

electrolysis, water 43
electro-osmosis 24–6, 58
electrodes
   care 46
   design for topical use 71–3
   intercalation 46
   platinum 43
   preparation 45–6
   silver/silver chloride 44–6, 101
electroincorporation 105–6
enkephalin 49–51, 80

factors affecting
   electroporation 100–2
   iontophoresis 18–24
fentanyl 30–1, 34, 102, 123

genes, delivery of 112–15
glucose extraction 26–8
GlucoWatch™ 27, 122

hair follicle density 34
heparin 102–3
histamine 65
human studies 123–4
hydrocortisone 48, 55, 59
hydrogels 51–5
hyperhidrosis, treatment of 63

inflammation, treatment of 58
insulin 22, 52–4, 90–4

Iomed, Inc. 61
ionic strength 23
iontohydrokinesis 24
implants for iontophoresis 9–10
IPPSF model 40–1, 60, 86
isoelectric point 78–9, 91, 93

ketoprofen 61–2, 124

lag time 7
LDF 129–30
Lectro Patch® 60, 70
LHRH and analogues
   leuprolide 25, 34, 36, 86–7
     LHRH 86, 102
     nafarelin 25, 36, 87
lidocaine 21, 33, 44, 58–61
liposomes 48–51

metoclopramide 123
metoprolol 29, 102
membranes, synthetic 36–7, 133
microdialysis 38–9
microiontophoresis 9
minoxidil 21, 68

nicotine 30
NSAIDs 61–3

ocular applications 9
oligonucleotide delivery of 110–12

parathyroid hormone 90
particles, delivery of 105–6
patch design 130–1
peptide transport 78–9
pH
   control and electrode material 43–5, 71, 73
   control by ion exchange resins 47
   formulation 23, 73, 78
phonophoresis 10–11
PHORESOR® device 58, 60, 69–70
pilocarpine 64
power supply
   electroporation 98–9
   iontophoresis 41

propranolol 25, 29, 39, 44, 51
protease inhibitors 80–1

regulatory issues 132–4
reverse electrodes 44
reverse iontophoresis 26–8

salt bridge 47–8
skin
   animal 33–4
   cultured 35
   electrical properties 13–14
   enzymatic activity 3–4
   entry of metal ions 44
   human cadaver 34–5
   immunology 4–5
   irritation 29, 123, 126–7
   maintaining viability 42
   pH 78
   proteolytic activity 79–80
   separation of epidermis 36
   stripping 43
   structure 2–3
   variability 124
stability of peptides 132
sufentanil 31, 34

tennis elbow, treatment of 59
TEWL 129
theoretical basis
   electroosmotic flow 26
   iontophoresis 16–19
   passive absorption 6
thyrotropin releasing hormone 20, 79, 83
transport
   efficiency 23
   pathways 5, 15–16
transport number 19, 24
transferosomes 49
TransQ® electrodes 58, 69, 72
treatment protocols 73–4

vasopressin 19, 52, 84–6
verapamil 29

water transference number 25